T0129570

The American Lifestyle in the 21st Century

Healthy Life, Happy Life

Marvin H Massey Sr. ND, PhD

Copyright © 2019 Marvin H Massey Sr. ND, PhD.

All rights reserved. No part of this book may be used or reproduced by any means, graphic, electronic, or mechanical, including photocopying, recording, taping or by any information storage retrieval system without the written permission of the author except in the case of brief quotations embodied in critical articles and reviews.

Balboa Press books may be ordered through booksellers or by contacting:

Balboa Press
A Division of Hay House
1663 Liberty Drive
Bloomington, IN 47403
www.balboapress.com
1 (877) 407-4847

Because of the dynamic nature of the Internet, any web addresses or links contained in this book may have changed since publication and may no longer be valid. The views expressed in this work are solely those of the author and do not necessarily reflect the views of the publisher, and the publisher hereby disclaims any responsibility for them.

The author of this book does not dispense medical advice or prescribe the use of any technique as a form of treatment for physical, emotional, or medical problems without the advice of a physician, either directly or indirectly. The intent of the author is only to offer information of a general nature to help you in your quest for emotional and spiritual well-being. In the event you use any of the information in this book for yourself, which is your constitutional right, the author and the publisher assume no responsibility for your actions.

Any people depicted in stock imagery provided by Getty Images are models, and such images are being used for illustrative purposes only. Certain stock imagery © Getty Images.

Print information available on the last page.

ISBN: 978-1-9822-2480-6 (sc)
ISBN: 978-1-9822-2481-3 (e)

Balboa Press rev. date: 04/05/2019

Contents

NES 1017 PhD Dissertation

The American Lifestyle and how we are destroying ourselves in the 21st Century.

A DISSERTATION

SUBMITTED TO THE GRADUATE SCHOOL

For the degree

DOCTOR OF PHILOSOPHY

Field of Natural Medicine
New Eden School of Natural Health
&
Herbal Studies

By
Marvin Massey, ND

McCordsville, Indiana

August 2018

Preface and Purpose

To teach people how to live a more healthy, happy, and successful life, I will endeavor to help people to change by example. When I change, the world changes. If I can teach two people how to live a better, healthier life, they in turn can share and teach two others and we can change the world by duplication.

Introduction

I heard a recent statement about our children, that the next generation (our children) will be one of the 1st generations not to outlive their parents.[1] Does that mean we should be prepared to bury our children? This should be a wakeup call, to all parents. Are we paying attention to all the junk we are feeding our kids? The processed, packaged, fast foods our children are eating are full of chemicals, preservatives and large amounts of sugar. The average child is ingesting about ½ pound of sugar daily. One of the biggest contributors to that statistic are carbonated soft drinks. These soft drinks are loaded with high fructose corn syrup. These foods and beverages all contribute to our obesity epidemic. The fake foods we are giving to our children are full of empty calories, simple carbohydrates, and antinutrients. It appears that the food giants are more concerned with the bottom line and their profits than with our health and the health of our children. The "standard American diet" (SAD) is full of pesticides chemicals and simple sugars. The current

---

[1] Dr Mark Hyman

population is overfed and mal-nourished; we have become one of the most obese nations on mother earth.

We are under way too much stress and there is never enough time to accomplish our daily tasks. We also rely on fast foods too much. By day's end we are exhausted and have nothing left in our tank. Therefore, we have no energy to do any kind of exercise which would help us lose some pounds and give us back some of the greatly needed energy that we are lacking.

We sit too much.[2] We cannot help ourselves if we do not know what to do or how to do it. One study indicated that our children are exposed to over 10,000 TV ads annually. More than half of these ads are glorifying sugary snacks, cereals and drinks.[3]

Additionally, they make fast food joints look like fun.

I would like to give the American people a sort of guide book or life style manual. We will use the basic principles of, nutrition, moderate exercise, attitude adjustment, service, gratitude, and overall life style changes. When each of us changes, the world changes. "Whatever the mind can conceive and believe it can achieve".[4]

We do not have the traditional nuclear family that used to exist 60 years ago where the family gathered around the table and ate whole foods together. Today we have many single parent families, most of whom are forced to work outside the home, and have little energy left at the end of a busy day to cook in a hot kitchen.

---

[2] Mayo Clinic

[3] Salt – sugar p80

[4] Think & Grow Rich p 32

Even those families who are fortunate enough to have two parents, both parents are many times working long hours and by day's end are very tired and stressed out.

Now is the time to take back control of your life, your health and the health of your children. All day every day we make choices to do this or that. To eat this or not eat this. We choose to embrace negative thoughts (ANTS)[5] or we do not. We can change our thoughts and our attitudes. Your thoughts matter. If you want to feel good, think good thoughts. Now is the time to change our priorities. What is important to you? Is it more money, a new car, trendy clothes, more stuff, or our health and wellbeing and the of health our children? Your response may be that I do not have children. OK, do you want more energy, do you want to sleep better at night? How would you like to begin to eliminate some stress from your daily routine? Then a few small changes are going to help you to accomplish your mission.[6] Sit down while eating, studies indicate that you will eat less sitting than standing. Slow down your eating and speed up your metabolism, taste your food, enjoy your food. Food is information for your body. Take a news fast from the TV and begin to eliminate some negative input into your life. Just try this for one week and see how you feel. Most of the news on our TV is bad news, earthquakes, floods, people being hurt and killed. It appears that most people watch the news at night just prior to going to bed. That means all the bad and negative news is rolling around in

[5] Change your Brain p 109

[6] Kaizen p 47

our minds while we are trying to get to sleep and stay asleep. Again, just take a news fast for only one week and see if you rest better and are more relaxed. The better your sleep, the more vitality and energy you will have.

We are an over medicated society; many of us are on multiple drugs. Too many people take drugs to go to sleep at night and drugs to wake up in the morning, and they rarely get a good night's sleep. We go through different cycles as we sleep. Early on we achieve most of the healing. If we do not sleep well, we retard our healing process.

I will endeavor to present to you some ways to change your lifestyle. We are currently on this collision course. As mentioned earlier, we sit too much. Get up and move! Walk, stand, and try to become more active. Do something; anything; just start moving on a regular basis. Walk down to the corner and back; you do not have to do extreme sports to receive benefits. Even gentle movements count, like tai chi and yoga. Just get up and move around. Standing is better than sitting, walking is better than standing and jogging is better than walking. The great thing about this is that you get to choose. Remember that something is better than nothing. When you do something good and healthy for yourself your sense of accomplishment will propel you forward to continue your journey. You will feel good about yourself and your motivation will carry you forward to your next goal. The important thing to remember is to start small, begin slow and build up to more. We must crawl before we walk, and we will have to walk before we can run.

Acknowledgements

I first, must thank my mentor and spiritual advisor, Fr. Justin Belitz OFM, and all the staff at the Franciscan Hermitage for all of the support and positive reinforcement I received prior to and during my studies.

I thank Professor George Morrell from the Univ. of Indianapolis for his guidance. For all of the proof reading and editing I thank Dr. Anna Sander DO and Dr. Kathleen Wiemold DC. Finally, I thank my wife Sharon, for all of the sacrifices she has made during the writing of this material, all of my love to you. Marvin Massey

Chapter 1

The White Menace

White sugar, white flour, white rice. For the most part, if the products we ingest, (aka foods) are of the white variety, it mostly indicates that the food has been bleached and processed. This would indicate that most of the nutrients have been processed out of this "food". The difference between feeling up or down, inspired or depressed, depends largely upon what we put into our mouth.[7] Completely eliminating sugar from your diet is a huge step toward better health and building a stronger immune system.[8] The number one addiction in America is sugar. Sugar exceeds cocaine, alcohol, or any drug! Sugar is a silent killer and it can cause all kinds of health issues that you may not even suspect. Sugars lack the vitamins and minerals required for their own metabolism and must draw on our body's stores of these nutrients. As these are depleted, our body becomes less able to carry out other

[7] Sugar blues pp 46

[8] Ultra-Metabolism p409

functions that require minerals and vitamins to be present. As a result, our cholesterol level rises, our metabolic rate goes down, fats burn more slowly.[9] We feel less like exercising, and we may become obese. Obesity increases the risk of diabetes, cardiovascular disease, and cancer. Cancer cells thrive on sugar and lack of oxygen. Refined sugar and simple carbohydrates comprise the basis for our Standard American Diet. Sugars feed candida yeast, fungi, other pathological organisms, and cancer cells.[10] Some of the more commonly known fungi are athlete's foot and toe nail fungus. These can both be treated with topical creams and ointments. What if we can cure them by eliminating sugar from our diet? Sugars interfere with the transport of vitamin C because both use the same transport system. Vitamin C's immune, collagen, elastin-building, and tissue glue forming functions are inhibited by sugar. The increased use of vitamin C and other supplements by 100 million Americans, against the advice of many physicians. Vitamin C is the king of the antioxidants (AO's). In our body. Vitamin E, C, carotene, bioflavonoids, and other AO's protect essential and other highly unsaturated fatty acids from free radical chain reactions. Vitamin C protects the watery parts of our body.[11] Sugars cross-link proteins, leading to wrinkles even in young skin.[12]

Sugar is poison; it is more lethal than opium and

[9] Fats that Heal p36

[10] Fats that Heal pp37

[11] Fats that heal p123

[12] Ibid pp37

more dangerous than atomic fall out. Now is the time for elimination. Heroin is nothing but a chemical, from the juice of the poppy, then it is refined into opium and then into morphine and finally into heroin.

Sugar also is nothing but a chemical. From the juice of the cane or the beet. Then it is refined into molasses, and after that refined into brown sugar and ultimately into the white crystals you see everywhere around you today.[13]

We must accept the fact that our bodies are hard wired for sweets. The old tongue map is incorrect. Research from the 1970's updates the tongue map; we now know that the entire mouth goes crazy for sugar. Even the palate. Scientists are now finding taste receptors all the way down our esophagus into our stomach and even into the pancreas, and they appear to be intricately tied to our appetites.[14] As soon as we see a sugary treat, the saliva in our mouth begins flowing, which prepares our digestive system. It is well known that digestion begins in the mouth. Not only sugar itself, but any refined starch, which the body converts into sugar. And it all begins in the mouth with the enzyme called amylase. The faster the starch is converted into sugar the quicker the brain gets the reward for it. The brain registers all forms of pleasure in the same way, whether it is a substance like a chemical drug like sugar or a satisfying meal. Among the many chemicals in the brain known as neurotransmitters are endorphins. These chemicals are very similar to morphine,

---

[13] Sugar blues p 22

[14] Sugar, Salt, Fat p 4

in that they tend to inhibit the communication of pain signals and produce a feeling of euphoria. Sugar seems to have that effect on most people.

Obesity is a global issue however, America is still the most obese nation on earth. In the twentieth-century America, obesity is being talked about and treated as a disease. Obesity is not a disease. It is only a symptom, a sign, a warning, that your body is out of order.[15] The data from 2008 shows that obesity among kids aged six to eleven jumped from 15 to 20%. Obesity equals illness. Chronic illness begins in our youth; it does not magically appear in our later years. We create most of our own ills due to diet, lifestyle, and negative thinking. Sugar is one of the bad guys, maybe the worst. Sugar has lost between 83 and 100% of each mineral present in raw sugar.

Once ingested sugar turns into saturated fats. British researcher John Yudkin blames sugar for the meteoric rise in cardio vascular disease (CVD). Sugar consumption is one of the quickest ways to increase TG's because our body turns sugar into fat to protect itself from the toxic effects of excess sugar. Increased use of these in our diet parallel the increase in (CVD) since 1900.[16]

---

[15] Sugar blues p 181

[16] Fats that Heal p 333

pH value of Blood should be between 7.35 & 7.45

Daily sugar intake produces a continuously over acid condition our body and requires more and more minerals to rectify the imbalance. Our body should be slightly more alkaline than acidic. Excess sugar is initially stored in the liver in the form of glucose (glycogen). Glucose is the sugar usually found with other sugars, in fruits and vegetables. It is a key material in the metabolism of all plants and animals. Glucose is always present in our bloodstream and is often called blood sugar.[17]

When the liver is filled to its maximum the excess glycogen is returned to the blood in the form of fatty acids which are stored in the belly, buttocks and thighs. When these nonvital areas are full, the fatty acids are then distributed among other organs such as the heart and kidneys. These begin to slow down and their tissues degenerate and turn to fat. The whole body is affected by their reduced ability and abnormal blood pressure is created. The parasympathetic part of the autonomic nervous system is affected. Thus, the circulatory and autonomic lymphatic systems are compromised, and the quality of red blood corpuscles are diminished which contributes to an overabundance of white cells and the creation of tissue becomes slower.[18]

---

[17] Sugar blues p147

[18] Sugar blues pp 137,138

Antibiotics

Dairy farms use antibiotics in cattle feed and injections. The egg industry adds antibiotics to feeds to keep cooped-up chickens healthy. Antibiotics encourage the growth of yeasts and fungi (including candida) in humans and can cause sugar craving (to feed candida) and hypoglycemia.[19] How many pounds of antibiotics are used annually? 32 million pounds are used on animals for prevention. 2 million pounds are used for cures on animals and 3 million pounds are used to cure humans according to Dr Mark Hyman. When we ingest beef or chicken we are eating what that animal has eaten.

Sugar can interfere with the proper assimilation of essential fatty acids (EFA's). Sugars, refined starches, commercial white flour, white flour pasta and white rice cannot support health and should be avoided. Whole wheat bread is not made with whole wheat. Whole wheat flour is white flour with a little bran added back, but it is still deficient in minerals, vitamins, EFA's and therefore cannot support our health.[20] Our soils have been depleted of many trace minerals. The plants cannot take out of the soil what is not there. If there are no minerals in the plants that the animals eat they will not have the EFA's in their meat. The word essential, (Essential Fatty Acids) in this case means that our body does not produce it and we must get it from an outside source.

[19] Fats that Heal p243

[20] Fats that Heal p318

Got Milk?

Lactose is milk sugar. Dairy is one of the major allergies in the Standard American Diet. Many people are allergic to milk and milk products and are not aware of it. We are one of few species that drink another species milk. Most animals are weaned from their mother's in their first year of life. The question that I have is that if the calcium in cow's milk is so beneficial for humans, why do we have the greatest incidence of osteoporosis of any nation on earth? One of the reasons is that we are not digesting and absorbing the calcium. Around age three our body stops producing lactase the enzyme that breaks down and digests lactose. What is cows milk designed to do? It is designed to take a 50 pound calf at birth and transform it into a 300pound cow in six months. Milk in its natural state, raw milk, from the cow's udder appears to be very healthy. If the cow grazes in fresh air on nutrient-dense, organic grasses, free range. This raw milk brought straight into our glasses would be rich in bioavailable nutrients and enzymes. Enzymes are the key to unlocking the usability of many nutrients.

Pasteurization

In the 1890's the farmers began pasteurizing milk due to pressure from the US government. This process destroys or greatly reduces the efficacy of enzymes and damages vitamins, proteins and probiotics. Pasteurization exposes raw milk to high heat or radiation for a certain amount of time, then cools it.

Homogenization is used to prevent separation of the cream from the milk. The cows are given BGH (Bovine Growth Hormone) to promote rapid growth and to increase the production of milk by 10 % to 15 %. First, they impregnate the cow to get the milk glands to start producing milk for her new calf. When the calf is born naturally the mother will want to nurse her baby. But they take the calf away and start milking her and of course, she has a lot of milk to give. After a few short weeks her milk will begin to dry up, then she is given BGH to stimulate her milk producing glands, so that her milk will not dry up. Then she is milked so frequently that her udder becomes infected. Now is the time for more antibiotics for the cow and pasteurization to kill the bacteria in the infected milk.[21]

More than 80 % of all antibiotics sold in the U.S. go to pigs, cows, chickens and other animals that people eat. The overuse of antibiotics has also created super strains of bacteria and viruses that are antibiotic resistant.

Sugar Names

Please note; "that anything that rhymes with gross is probably a sugar." Fructose is fruit sugar, in normal amounts your body assimilates well. Maltose is malt sugar. Lactose is milk sugar, not digested by most people past

[21] Bring back vitality pp72,73

age three. Glucose, as mentioned earlier, is always present in our blood stream. There are two sugars that are of greatest concern. Sucrose is the refined sugar made from the sugar cane and the sugar beet. Dextrose is derived synthetically from starch also called "corn starch".[22]

High fructose corn syrup, which is now labeled just corn syrup is such a menace because it is in liquid form and cheaply added to the beverages that are on the market. In 1962 the per capita consumption of soft drinks was 16.2 gallons, by 1970 that number almost doubled and climbed to more than 30 gallons per person.[23]

Tooth Decay

Loma Linda School of Dentistry, two researchers found that subtle changes in the internal activity of the teeth caused by sugar can be an early sign of later decay. A high sugar diet can slow the rate of transport of hormonal chemicals by as much as two-thirds. Teeth with sluggish internal activity have a high incidence of decay. A hormone released by the hypothalamus stimulates the release by the salivary or parotid gland of a second hormone. This hormone increases the rate of fluid flow in the teeth. A high sugar diet upsets the hormonal balance and reduces the flow in the internal system. This weakens the tooth and makes it more susceptible to decay. Healthy teeth are normally invulnerable to the microbes

---

[22] Sugar blues p 147

[23] Sugar blues p 179

in the mouth.[24] Physical degeneration on a global scale is related to sugar consumption. Dr A. Kawahata, a leading Japanese nutritionist from Kyoto University, quotes an early Buddhist axiom: If you look for sweetness – Your search will be endless – You will never be satisfied – But if you seek the true taste – You will find what you are looking for.

Dr Weston A Price, dentist from Cleveland Ohio, traveled all over the world in 1935. He published a book on his findings in 1939. In area after area he visited the so-called backward primitive conditions had excellent teeth and wonderful general health. They ate natural, unrefined food from their own locale. As soon as they were introduced to civilized, refined, sugary foods, physical degeneration began in a way, that was observable, within a single generation.[25]

As Dr Price continued his travels, the results were the same, there was an obvious and direct relationship between the increase of tooth decay and the adoption of a modern diet. Additionally, he found deformities in the facial bones, which caused crowding of the teeth. These malformations of face and head were correlated with lower IQ's and personality disturbances. Along with a higher incidence of degenerative diseases, such as tuberculosis.[26]

Where ever he went, Dr Price found that beautiful straight teeth free from decay, good physiques, a general resistance to disease, and fine characters were

---

[24] Sugar blues pp198,199

[25] Ibid p142

[26] Diet & Nutrition p20

typical of native groups on the traditional diets, rich in essential nutrients. The remote peoples Price initially photographed were noteworthy for "their fine bodies, ease of reproduction, emotional stability, and freedom from degenerative diseases typical of the civilized moderns subsisting on the displacing foods of modern commerce, including sugar, white flour, pasteurized milk, low-fat foods, vegetable oils, and convenience items filled with extenders and additives."

The discoveries, theories, and conclusions of Dr Price are noted in his classic work, *Nutrition and Physical Degeneration.* Dr Price documented the classical characteristics of traditional diets. The diets of healthy, non-industrialized people contain little or no refined sugar, white flour, canned foods, low-fat milk, hydrogenated vegetable oils, protein powers, synthetic vitamins or artificial colorings.[27]

White Rice

Throughout recorded history, whole grain products have been dietary staples of agrarian societies. Polishing rice removes a large portion of many minerals and vitamins, especially the B vitamins. Only about 60% of the riboflavin remains in polished rice, only one-third of the niacin and less than one-half the pyridoxine. This situation is even worse with thiamine, or vitamin B1, where only 20% remains in the rice. In cultures where

[27] Epigenetics p118

the people subsist almost exclusively on polished rice this lack of thiamine can result in a disease called Beri-Beri. [28] Polished rice has lost between 26% and 83% of each mineral present in brown rice.[29] We generally associate rice with China. The Chinese written character for food is the same as that for rice. However, rice originated in India. It is said that Buddhism, when it spread from the sub-continent to the Far East, brought with the custom of eating rice.[30]

Carbohydrates

There are two types of carbohydrates; simple and complex. Simple carbs are sucrose. Complex carbs are vegetables, grains and a little fruit, these carbs are the best sources of slowly released glucose, which is the best fuel for providing the energy we live on. Complex carbs contain fiber and other materials that are digested slowly. Their starches are slowly converted into glucose, which is then burned (oxidized) in body functions at the same rate at which it is produced. Complex carbs do not produce excess energy that turns into fat. Complex carbs also contain vitamins and minerals (co-factors) that enable our body to burn them cleanly into carbon dioxide, water, and energy.[31] Refined or simple carbs contain no protein, no

---

[28] Diet & Nutrition p76

[29] Fats that heal p76

[30] Ibid p75

[31] Fats that heal p33

minerals, and no fat or fiber. Simple carbs are essentially empty calories. Our body can turn excess sugar into fat, but it cannot turn fat back into sugars. It must burn off the fat through activity. An important note here is 65% of the white sugar available commercially is made from GMO, (genetically modified) sugar beets. Please avoid at all costs.

Some History on Sugar

From the Garden of Eden through thousands of years, what we call sugar was unknown to mankind. We evolved and survived without it. None of the ancient books made mention of it, Mosaic Law, I Ching, The Yellow Emperors Classic of Internal Medicine. The Greeks had no word for it. In 325 B.C. Nearchus, one of Alexander the Greats Admirals described it as a kind of honey. The Greek Herodotus called it manufactured honey and the Roman Pliny called it honey from the cane. It was used like honey, as a medicine. Dioseoridrs, A Roman writer of Nero's time described it as a sort of concreted honey, which is called saccarum found in cane in India and Arabia Fleix.[32]The University of Djondispour of Persia first refined and solidified the juice of the cane into solid form around 600 A.D. The original Sanskrit word for morsel or piece was attached to this "Indian Salt" The armies of Islam overtook the Persian Empire and one of

[32] Ibid p28

"Spoils of War" was the secret of processing sweet cane into medicine.

An early European observer, Rauwolf, credits the widespread use of sugar by Arab desert fighters as the main reason for the loss of their sharpness in battle. In 1573, he wrote about the Turks and Moors. The Turks gorged themselves on the sugar cane and became less courageous and no longer the intrepid fighters they had once been.[33] Moving ahead in history, in the 1960's, High Fructose Corn Syrup (HFCS) had higher levels of fructose than of glucose. It then became very popular in the 1980's, because of its long shelf life that processed foods demand. On a scale of sweeteners sucrose comes in as 100, glucose comes in at 74, and fructose registers a 173.[34] It should not come as a surprise that the introduction of HFCS into our food supply is associated with the beginning of the obesity epidemic.

The above cited historical events and results point to many of the problems caused by sugar. It appears that sugar can affect us mentally, physically and emotionally. In completing the first step of our journey to a healthier lifestyle it becomes evident that if we avoid sugar and its many forms we will move ourselves in the direction of health and happiness.

---

[33] Ibid p 30

[34] [34] Salt, Sugar p 130

Chapter 2

Farm to Factory – Organic – Food as Medicine

Our ancestors did not go to the local grocery store to shop, and then find the organic section, to be sure that they were getting the very best possible food available. Less than 100 years ago a great number of the population were farmers and the produce was full of vitamins and minerals.

Through culture evolution, human beings changed their environment in unprecedented ways, even as they adapted to its demands. By domestication of animals, and the mastery of agricultural practices humans transitioned to new means of food production through farming and animal husbandry, this transition in known as the Neolithic Revolution.[35] New disease patterns became a part of the price paid for living in large, densely populated, permanent towns and cities. Endemic diseases (that which is restricted to a peculiar locality or region) and epidemic

[35] History of Medicine p 2

diseases (affecting large numbers of individuals within a population or community at the same time) which may determine the density of populations, the dispersion of people and the diffusion of genes, as well as the success or failure of battle and colonization.

Recent studies of the origin of agriculture suggest that it was almost universally adopted between 10,000 and 2000 years ago primarily in response to pressures generated by the growth of the human population.[36]

Historians and archaeologists have revealed that ancient civilizations all over the world recognized that certain foods could provide health promoting and disease-protective benefits. Food supplies us not only with the basic nutritive functions, but also with a secondary level of nutrition that adds a complex layer of disease resistance and longevity benefits.[37] Mesopotamia, Egypt, China and India and Greco-Roman are considered the four earliest civilizations. In contrast to the gradual evolution found in Egyptian, Indian and Chinese history, Greek civilization seems to have emerged suddenly. Whatever their origins, the intellectual traditions established in ancient Greece provided the foundations of Western philosophy, science and medicine.

---

[36] Ibid

[37] Super Immunity pp11, 12

Ancient Egyptian Medicines

Food was needed to sustain life, but as it passed through the intestinal tract it was subject to the same putrid process that could be observed in rotting foods, wounds, and un-embalmed corpses. If the products of decay were strictly confined to the intestines, eating would not be so dangerous, but putrid intestinal minerals often, contaminated the system of channels that carried blood, mucus, urine, semen, water, tears, and air throughout the body, causing localized lesions and systemic diseases.[38]

In contrast to Mesopotamian custom, Egyptian prescriptions were precise about quantities.

It does not take years of study and contemplation to become an expert in human nutrition, so long as you understand the principals that govern your basic food choices and food preparation. When we eat healthy foods, we become healthy, if we do not, we will develop diseases. Essentially, we are made from the foods we eat. A common statement is that "we are what we eat".[39] However, I think it goes deeper that that. I believe that we are what we digest and assimilate. Please note, that currently the average American takes over 60% of their calories from processed foods, this number has increased gradually but steadily over the last hundred years.

---

[38] History of Medicine p30

[39] Super Immunity p13

India

India is densely populated with a mixture of races, languages, cultures and religions. The subcontinent is a world of bewildering complexity. The native healing art of India is Ayurveda, the learned system that forms the basis of the traditional medicine widely practiced in India today, it is known as "The Science of Life". There are four pillars of Ayurvedic Medicine, the physician, the medicine, the attendant and the patient. The primary objective of the Science of Life was maintenance of health, rather than treatment of disease.[40] Prevention rather than cure. "Lifestyle"

One of the methods used by the physician was the assessment of the odor and taste of secretions and discharges. The most famous diagnostic taste text was for the "honey urine disease" (diabetes).[41]

Traditional Chinese Medicine TCM

Classical Chinese medicine is based on the belief in the unity of nature, the Yin-Yang dualism, a medical practice based on the theory of systematic correspondences and like the Indian Ayurvedic Medicine used the theory of five phases or elements.[42] During his 100 year-reign, Huang Ti and his Minister of Health and Healing Ch'i

[40] History of Medicine p 41
[41] Ibid p 43
[42] Ibid p 47

Po compiled the Nei Ching (The Yellow Emperor's Classic of Internal Medicine) which has guided Chinese medical thought for over 2500 years.

The Chinese anatomy is concerned with the dynamic interplay of "functional systems" rather than organs. The emphasis is on function rather than structures. The Chinese use the concept of these ceaseless, circular movements of blood and energy within the body's networks of channels referred to as meridians. The 365 energy points on these meridian channels are used by acupuncturists. Using the teachings of Huang Ti, the sages of ancient times did not treat those who were already ill. Instead they gave the benefits of their instruction to those who were healthy. The scholar practiced preventive medicine and took no fee for his work. According to the Nei Ching, a diet balanced in accordance with the five-fold system of correspondences, would promote health and longevity, strengthen the body and drive out disease. Therefore, all therapies were directed towards restoration of the state of harmony.[43]

Harmony = Balance = Homeostasis.
Prevention over cure. "Lifestyle"

Greek Medicine

Shamanistic, religious, and empirical approaches to healing are as we have seen, universal aspects of the history of medicine. Greek medicine appears to be unique

---

43 History of Med p 51

in the body of medical theory associated with natural philosophy. The earliest Greek natural philosophers were profoundly interested in the natural world. By the sixth century B.C. Greek philosophers were attempting to explain the working of the universe in terms of everyday experience. Pythagoras 530 B.C. is said to have been the first Greek philosopher with a special interest in medial subjects. The Pythagorean, approach was apparently inspired by mathematical inquiries. All things could be divided into pairs. The harmony of balancing pairs, such as hot and cold, moist and dry was especially important in matters of health and disease.[44]

Hippocrates the "Father of Medicine" 460 – 361 B.C. Plato and Aristotle speak of Hippocrates with respect, despite the fact, that he taught medicine for a fee. The Hippocratic moto is "at least do no harm". In the text written by Hippocrates "On Ancient Medicine". A major thesis of this work is that nature itself has strong healing forces. The purpose of the physician, therefore, was to cultivate techniques that would work in harmonious balance. Thus, to care for his patient, the physician must understand the individuals, constitution and determine how health was related to food, drink, and mode of life[45] "Lifestyle"

In a fundamental sense, dietetics was the basis of the art of healing. The microcosm or small world of the human body reflects the four elements, earth, air, fire and water. The Hippocratic theory is that health is the result of

[44] Ibid p 65

[45] History of Med p 68

the harmonious balance and blending of the four humors, which parallel the four elements.[46]

For medical science, the Hellenistic period, about the time between the death of Alexander the Great to around 30 B.C., was when the Romans annexed Egypt. Most notable about this time frame was the work of Herophilis and Erasistratus, who were both anatomists. For Herophilis, health was the greatest good. He is credited with making this statement "Wisdom and Art, Strength and Wealth, are useless without Health". Like Hippocrates, Erasistratus (310 250 B.C.) was born into a medical family. He saw the body tissues as a network of veins, arteries, and nerves. Erasistratus reasoned that although we can consciously conceal our thoughts, their influence on the body cannot be controlled. This is very similar to, the emotional overload we sometime cannot control. Negative thoughts can make us ill.

Medicine in the Roman World

Cato the Elder (234 – 149 B.C.), denounced Greek physicians. His favorite traditional remedy was cabbage, which in addition to being harmless, it was a good source of vitamin "C".

Pliny the Elder (23 -79 A.D.) like Cato before him, was suspicious of professional physicians. In Pliny's "Natural History", he claimed that everything had been created for the sake of mankind, making the whole world

[46] Ibid p 71

an apothecary for those who understood nature's simple prescriptions.

Dioscorides (40 -80 A.D.) compiled the "Materials of Medicine", one of the first Western herbals. An acute naturalist, he provided valuable information about medically useful plants. Plant food as medicine. Many of the herbal remedies identified and classified by Dioscorides can be found on our modern spice racks. For example, cinnamon and cassia were said to be valuable in the treatment of internal inflammations.[47]

In providing an historical context for his discussion of medicine, Celsus noted that the Greeks had cultivated the art of medicine more than any other people. The ancient Romans had enjoyed natural health, without medicine, because of their good habits and lifestyle. When they turned to lives of indolence and luxury, illness appeared and medicine become, a necessity. Celsus considered it essential for every individual to acquire an understanding of the relationship between disease and the stages of life. Acute illnesses were the greatest among the young and chronic illnesses threatened the old. Celsus' conclusion that the best prescriptions for a healthy life was one of variety and balance, proper rest and exercise.

Galen (130-200 A.D.) No known figure in the history of medicine has influenced concepts of anatomy, physiology, therapeutics, and philosophy as much as Galen, the physician known as the "Medical Pope of the Middle Ages", and the mentor of Renaissance anatomists and physiologists. He originally arrived in Rome in 161

---

[47] History of Med p 83

A.D. and his patron the Emperor Marcus Aurelius called him the "First of Physicians and Philosophers." According to Galen, the best physician was also a philosopher, and must master three branches of philosophy: logic, physics and ethics. Galen assumed that nature acts with perfect wisdom and does nothing in vein, he argued that every structure was crafted for its proper function. Galen was honored as the ultimate authority on anatomical and physiological questions without serious challenge until the sixteenth century. Galen established the foundations of a program that would transform the Hippocratic Art of Medicine into the Science of Medicine. [48]

This brief history of ancient medicine contains a common thread or theory that the body is created to function properly if given the proper nutrition, exercise, rest and with the correct mental attitude. By following these simple guide lines, we can achieve harmony and balance which will aid us in changing our lifestyle for the better. We can change, with diet and lifestyle we can cure chronic diseases. Some of the underlying thoughts are also inclusive of service to others without being paid.

Dr. Joel Wallach says "Give yourself the best nutrition you can, and let the healing be done by the greatest physician of all ..... Your body."

---

[48] 48 Medical History pp 86-90

Nutrition and the Soil

The quality of human life has remained, to a large degree, dependent on the quality of the interaction between dust, plant matter and the microorganisms of the soil.

Unfortunately, over the Millenia soils in many areas have been depleted and destroyed. On September 4th, 1882 Thomas A. Edison pulled the switch, in New York, that started the first commercial electric generating plant in the world. Electricity satisfied many needs of mankind like, fuel for heating, cooking and lighting.

However, most people do not consider the down side for or soil. Prior to electricity we used wood for heating and cooking. The byproduct of wood burning is ash. When you burn wood, you create ashes. Ashes contain all the minerals from the wood and was normally recycled back into the soil when used for gardening and farming. Without ashes as a source of nutritional minerals, agriculture and livestock suffered. Many farmers went out of business without the serious supplementation of dietary minerals.[49]

Because of intensive farming, poor crop management, erosion, commercial fertilization, the use of pesticides, and other problematic factors much of the soil in which our crops are now raised has been depleted, particularly of essential minerals.[50]

The best farmers replenish the soil as it is farmed.

---

[49] Epigenetics p 120

[50] Naturopathy p 195

Unfortunately, this practice has become the exception rather than the rule. The most direct and immediate losses are the mineral and vitamin deficiencies in the soil that passes up the food chain to humans.[51]

In the 1940's when transportation of food stuffs in the United States was still rather limited, what most people ate was grown in the region where they lived.[52] Soil depletion and the use of pesticides and chemical fertilizers result in food that is less safe and less nourishing. Soils in many parts of the world are deficient in certain minerals. This can result in low concentration of major or trace minerals in drinking water and plant crops.[53]

Research in Europe and in America has shown that farms using no chemical fertilizers, pesticides or herbicides have consistently turned out yields as good as their conventional counterparts and in many cases at less expense. In China, for example, a study of the agricultural practices has revealed that composting was a highly developed art. Every fragment of material that came from the land was returned to it. This included not only the animal manure, but also the plant wastes. All these components were carefully mixed into heaps which would encourage the growth of desirable bacteria and fungi, and the finished material was then returned to the land.[54]

The soils of the world have suffered at the hands

---

[51] Ibid p 96

[52] Diet and Nutrition p 24

[53] Naturopathy p 17

[54] Diet & Nutrition 33-36

of farming. The present food production system, while correcting some abuses of the past, still inflects on the soil a variety of new and old insults that diminish its nutrient value. Ecological health depends on keeping the surface of the earth rich in minerals and humus so that it can provide a foundation for healthy plant and animal life. Our dependence on artificial manmade products interferes with our relationship with the soil and the natural world in general. Because of this, nutritional supplementation is necessary.[55]

---

[55] Naturopathy p 195

Chapter 3

Supplements

Ultimately, we are what is in (or is not in) the food we eat, and what we absorb.[56]

As we track the evolution of human thoughts and beliefs as they relate to science and health, the path from belief in disease caused by evil spirits to beliefs in alchemy, witchcraft, sorcerers, physicians, the germ theory, genetics, the elucidation of the double helix, the mapping of the genome and finally to the belief in epigenetics has taken thousands of years.

Epigenetics states that the gene and the genetic code is not a blue print but is instead a "script" that is easily modified, interpreted, or differentiated by actors, directors, and producers. The DNA is the script.[57] DNA can be altered by the diet.

There are approximately 90 essential nutrients we need on, a daily basis. America is not calorie deficient, but

[56] Epigenetics

[57] Ibid p 208

we are malnourished because of a lack of the 90, essential macro-and micro nutritionals. These mineral deficiencies are some of the direct causes of America being the most obese nation in the world. America's great interest in fast foods, fad foods, snack foods, desserts, caffeine, alcohol, drugs, and tobacco, is a symptom and a direct result of mineral deficiencies.

There are seventy-five metals, listed in the periodic table, all of which have been detected in our blood and other body fluids. We know that at least sixty of the metals (minerals) have a direct or indirect physiological value for animals and humans.

Organically not a single function in the animal or human body can take place without at least one mineral or metal cofactor.[58] Whether we realize it or not minerals have a governing effect on our lives. Ninety two percent of Americans are deficient in one or more of the essential vitamins and minerals. More than ninety nine percent of Americans are deficient in the essential omega- 3 fatty acids.

Some of our immediate concerns should be the nutritional deficiencies and the self-destruction inflicted by eating too many fried and over cooked (burnt) foods, by the overuse of dietary oils (including margarines), salad dressings, cooking oils, as well as the consumption of canned fish packed in oils, and processed meats preserved with nitrates and nitrites. These items all, have a tendency to increase the daily load of dietary trans fatty acids.[59]

---

[58] Ibid p 236

[59] Ibid p 288

Vitamins will not function without minerals. No plant, animal, or human can produce these 60 essential minerals; they must be consumed. We cannot get the essential nutrients from our foods if they are not in our foods. The nutrients will not be present in our foods if they are not present in the soil for our bodies to absorb, and the best way that we can get the nutrients is in plant form. It is easy to ascertain that if the plants are deficient, it is because the soil they grow in is deficient in minerals.

It is likely that plant food consumption throughout much of human evolution shaped the dietary requirements of contemporary humans. Diets would have been high in dietary fiber, vegetable protein, plant sterols and associated phytochemicals, and low in saturated fats and trans-fatty acids and other substrates for cholesterol biosynthesis.

Clearly our ancestors did not take supplements. However, they also ate 100% organic food, grown in nutrient rich soils, drank only pure water, and ate no processed foods with additives and pesticides. They had clean air and exercised every day. It is not possible to eat a pure, and sufficient diet, in our industrialized world. Soils are nutrient depleted and so is our produce, even the organic produce. Most of us do not have the time or desire to eat the amount of raw vegetables required to match the vitamin, mineral, and fiber content of the diets of our ancestors. We do not exercise by walking around all day hunting and gathering. The reality is that supplementation is now a requirement in order to achieve the nutrient and fiber levels that our cells require.[60]

---

[60] Innate diet pp 150-151

Minerals

Minerals are the currency of life. The basic functions of life itself cannot be preformed without optimal amounts of minerals. None of our biologic processes are exempt. Minerals can affect vitamin function, hormone functions, and act as a cofactor to facilitate a chemical or enzymatic reaction. Minerals and mineral supplements to our diet are necessary.[61]

Dr. Gerhard Schrauzer, the Chair of the Department of Chemistry UCSD, was instrumental in the research on the mineral selenium, showing that it was an essential nutrient. Dr. Schrauzer is to selenium what Dr. Linus Pauling is to vitamin "C"[62]

It is fairly well understood that certain "Genetic Diseases", like type 2 diabetes, can be cured with certain minerals, along with lifestyle changes.

The US Senate Document 264 states; there are few if any nutritional minerals left in our farm and range soils thus, there are few if any nutritional minerals left in our grains, nuts, fruits or vegetables. It is necessary to supplement all of the major minerals, trace minerals, and rare earth minerals.[63]

Statistics for the 2005-2006 USDA, show that the average dietary intake of the most commonly measured nutrients are below the RDA (recommended daily allowance). The mineral depletion of the soil has come

---

[61] Epigenetics pp 288-289

[62] Ibid p 319

[63] Ibid p 403

about through poor soil management. Too many chemical fertilizers substituting for composted organic matter has left our soils vulnerable to erosion and loss of many of our minerals. The plants and crops grown on the soils are not healthy nor as rich in nutritional value as those grown on healthy soils.[64] It is interesting to note that city dwellers may require more supplements than country folk. It is understandable that the city life is more hectic and chaotic than the country life which is normally a quieter atmosphere.

As always, we should use common sense about our supplementation.

Minerals in pill form are, after all, only adjuncts to a diet. They are supplements not substitutes. There are three basic forms of minerals; colloidal, chelated, and metallic. It is fairly, common knowledge that in cultures with the longest living people, it comes down to the availability of the source of essential nutrients, on a daily basis. Some of the most important elements are the plant-derived colloidal minerals.

Colloidal minerals: a colloid refers to a substance that exists as ultra-fine particles that are suspended in a medium of different matter. The solutions part of a colloid provides a solid, gas, or liquid medium in which the colloid particles are suspended.[65]

The molecular groups of particles of the colloid solute carry a resultant electrical Charge. Generally, all of the particles of these inorganic colloids carry the same

---

[64] g Ibid

[65] Epigenetics p 403

negative sign. A small percent of these inorganic colloids will pass through the intestines of a living animal or human, because a natural chelating process takes place in the gut in the presence of protein containing foods.[66]

Chelated minerals were created in the 1960's by the livestock industry. The term chelated means "claw" in the Greek language. This is used to describe the process by which an amino acid, protein, or enzyme, is wrapped around the mineral atom, alloy or molecule. This enhances the bio-availability of the metallic minerals. Metallic minerals include egg shells, oyster shells, calcium carbonate, lime stone, dolomite, clay, mineral salt, seawater, Great Salt Lake Waters, minerals oxides, vortex water, sea bed clay, clay soils, and "rock flours". [67]

Despite all, of the claims, metallic minerals are absorbed only eight percent to twelve percent in all vertebrates, including humans. It is important to note that after the age of 35 to 40 years of age, the absorption rate available in humans is reduced to the point where some people only absorb around three to five percent.

Essential minerals function at the subcellular level as cofactors for the optimal operation of genes, DNA, RNA, chromosomes, enzymes, vitamins, and hormones in all organisms, plant and animals including humans.

All enzyme activity involves minerals. Minerals are essential for the proper utilization of vitamins and other nutrients. The level of each mineral in the body has an

---

[66] Ibid p 404

[67] Ibid p 402

effect, on every other mineral, so if one is out of balance, all mineral levels are affected.

Nutritionally, minerals belong to two groups; bulk minerals (macro-minerals) and trace minerals (micro-minerals). Bulk minerals include calcium, magnesium, sodium, potassium, and phosphorus. These are needed in larger amounts than trace minerals. Although only minute quantities of trace minerals are needed, they are still important for good health. Some trace minerals would include boron, chromium, copper, germanium, iodine, iron, manganese, molybdenum, selenium, silicon, sulfur, vanadium, and zinc.[68]

Mineral supplementation can help you make sure you are getting all the minerals your body requires. When mineral supplements are taken with a meal, they are automatically chelated in the stomach during digestion. The absorption of minerals can also be affected by the use of fiber supplements. Fiber decreases the body's absorption of minerals. If you supplement with fiber and minerals, you should take them at different times.[69]

Our body contains more calcium than any other mineral, and the calcium is mostly contained in our bones. Calcium aids in the prevention of osteoporosis.

Recent research has shown that magnesium deficiency may contribute to the formation of kidney stones. Magnesium aids in maintaining the body's proper pH balance and normal body temperature.

Selenium is a vital antioxidant, especially when

---

[68] Nutritional Healing 2010 p 32

[69] Ibid p 33

combined with vitamin E. It also strengthens the immune system by preventing the formation of free radicals. Selenium with vitamin E and zinc may also provide relief from an enlarged prostate. Selenium deficiency has also been associated with cancer and heart disease.[70]

I chose the afore mentioned three minerals because they are in the news often and there are, a number, of studies showing that as a nation, America is deficient in these minerals. As a nation we are deficient in many, minerals and I will focus on these.

On a, daily basis, we should try to ingest 60 minerals. Liquid is best, followed by gel caps, then capsules. Tablets or coated pills seem to be the most difficult for humans to digest. I personally do not endorse gummies. Read the labels, many of them use sugars to entice children to take them, this sort of defeats the purpose of the supplement, whether it is a vitamin or mineral.

As always research the product and use the best quality product available. You and your family are worth it.

Vitamins

Vitamins: we will identify 16 vitamins that are needed on a regular basis. The definition of a vitamin is: any group of organic compounds which are essential for normal growth and nutrition and are required in small quantities in the diet because they cannot be synthesized by the body (A serving containing a specified amount

70 15 Ibid p 40,

of a particular vitamin, or vitamins, taken as a dietary supplement). Origin of the word vitamin: early 20th century based on the Latin vita "life" + amine meaning amino acid, because vitamins were originally thought to contain an amino acid.[71]

The 16 Vitamins are A, B1, B2, B3, B5, B6, B12, C, D, E, K, Biotin, Choline, Flavonoids, Folic Acid, and Inositol. There are three categories of vitamins: Fat soluble, water soluble, and flavonoids.

Fat soluble vitamins cannot be efficiently absorbed from the intestine when humans consume low fat diets. Vitamins A, D, E, and K are fat soluble vitamins and require fat to be absorbed.

Vitamin A is an antioxidant and necessary for good vision, healthy skin, bones, and teeth. Carotenoids are closely related to vitamin A. The best known carotenoid is beta carotene, which is converted into vitamin A in the liver. B vitamins act as coenzymes and are involved in energy production and work together. The B vitamins are responsible for maintenance of healthy nerves and proper brain function. B1 (thiamine) enhances circulation, carbohydrate metabolism, and in the production of hydrochloric acid in the stomach. B2 (riboflavin) is necessary for red cell formation, anti-body protection, cell respiration and growth. B3 (niacin, nicotinic acid, niacinamide) is a coenzyme, it interacts with cellular respiratory enzymes and is essential for the oxidation reduction reactions in the release of energy from carbohydrates, fats, and proteins.

---

[71] Wikipedia

B5 (pantothenic acid) was isolated in 1940. One of its primary roles is as a componentof the coenzyme A molecule. It contributes to some reactions which are essential for the release of energy from carbohydrates in gluconeogenesis, and in the synthesis and degradation of fatty acids. It is also needed in the synthesis of essential compounds such as sterols and steroid hormones. Also known as the anti-stress vitamin, it plays a role in the production of adrenal hormones and the formation of anti-bodies.

B6 (pyridoxine), is an essential coenzyme for glycogen phosphorylase. Pyridoxine isinvolved in more bodily functions than any other single nutrient. And it has the ability, to affect both physical and mental health. It is necessary for the production, of hydrochloric acid and the absorption of fats and proteins. It aids in maintaining the sodium and potassium balance and promotes red blood cell formation. Pyridoxine is required by the nervous system and is necessary for normal brain function and for the synthesis of nucleic acids RNA and DNA.

B12 (cyanocobalamin / methyl-cobalamin) is the most chemically complex of all the vitamins and is the general name for a group of essential biological compounds known as cobalamins. The liver does convert a small amount cyanocobalamin into methyl-cobalamin. The term Vitamin B12 and cobalamin are used for all of the cobalt-containing corrinoids that can be converted to methyl-cobalamin and become active in human metabolism. B12 deficiency caused by malabsorption is most common in elderly people. B12 in the methyl-cobalamin form, may help prevent Parkinson's disease and slow the progression in those who already have the disease. Methyl-cobalamin is essential in converting homocysteine into methionine, which is used to build protein.[72]

B12 is required to support the function of bone marrow and the production of the myelin sheath, the nerve fiber insulation coating, and plays a key role in folic acid metabolism.[73] Biotin: in the 1930's there was a condition called the "egg white syndrome". In 1940 Paul Gyorgy identified the substance to cure this condition as biotin. Biotin is required for glucose metabolism, and for, the production of fatty acids. Biotin is a sulfur containing vitamin. Biotin is needed, in sufficient quantities, for hair growth and skin as well as prompting healthy sweat glands, nerve tissue, and bone marrow. Additionally, it also helps to relive muscle cramps.

Folic Acid, folate, folacin: identified in 1946, folic acid is required for the synthesis of DNA, RNA, and erythrocytes. Folate and Folacin are generic terms for compounds that have nutritional activity and chemical structures like those of folic acid.[74]

Folate is considered a brain food, and is needed for energy production and the formation of red blood cells. It also strengthens immunity by aiding in the proper formation and function of white blood cells. Because it functions as a coenzyme in DNA and RNA synthesis, it is important for healthy cell division and replication.

---

[72] Nutritional Healing p 23

[73] Epigenetics p 387

[74] Ibid p 388

Folate works best when combined with vitamin B12 and vitamin C.[75] Some medications can interfere with folic acid absorption and metabolism for example, aspirin and oral contraceptives.

Choline is a major structural component of larger molecules, as a component of phosphatidylcholine (lecithin), it is essential to the structure of all cell membranes, plasma lipoproteins, and pulmonary surfactants. Choline is needed for the proper transmission of nerve impulses from the brain through the central nervous system, as well as for gallbladder regulation, liver function, and lecithin formation. Choline is beneficial for disorders of the nervous system. A high choline intake can keep inflammation in the body low, which can reduce the risk of heart disease.[76]

Inositol, aka myo-inositol, is a cyclic alcohol that is chemically, similar, to glucose. In plants it is phytic acid and in animals myo-inositol it is a major constituent of phospholipids in bio-membranes. Myo-inositol is believed to be an essential nutrient because myo-inositol triphosphate is a second messenger for receptor mediated hormonal stimuli for mobilizing intracellular calcium.[77]

Inositol is vital for hair growth. This vitamin has a calming effect and helps to reduce cholesterol levels. It helps prevent hardening of the arteries and is important in the formation of lecithin and the metabolism of fat and cholesterol. It helps remove fats from the liver.

---

[75] Nutritional Healing p 24

[76] Ibid

[77] Epigenetics p 392

Research indicates that high doses of inositol may help in the treatment of depression, obsessive-compulsive disorder, and anxiety disorders, without the side effects of prescription medication.[78]

Vitamins are micro-nutrients and the body only requires them in small amounts compared to such nutrients as carbohydrates, proteins, fats and water. Vitamins are a collection of unrelated organic compounds that are necessary cofactors for metabolic chemical reactions within cells and essential for normal growth and maintenance of health.

Vitamins are essential nutrients and most of them need to be manufactured in the Body, many of which perform as coenzymes. Vitamins do not supply calories or contribute to body mass. Vitamins regulate metabolism, participate in the citric acid cycle and conversion of fat, sugar, carbohydrates, and proteins into energy.[79] Oil soluble vitamins are also referred to as fat soluble vitamins. For our purposes these terms will be used interchangeably. Water soluble vitamins, like C and B complex vitamins cannot be stored in the body, as they are excreted within four hours to one day. Water soluble vitamins should be taken daily.

Oil based vitamins should be taken when needed In small amounts, they will over time be stored in body fat tissues and the liver. Oil based vitamins include A, D, E, and K.[80] Taking the vitamin and mineral supplement with

---

78 Nutritional Healing p 25

79 Epigenetics p 373 -374

80 Nutritional Health p 18

meals helps to assure a supply of other nutrients needed for better assimilation. In nature, there are no stand-alone nutrients. All of nature's nutrients work together for the common cause of balance.

The author of Nutritional Healing, Phyllis A. Balch, recommends taking the oil soluble vitamins before meals, and the water-soluble vitamins after meals.[81] As stated before, nature's nutrients work together. For example, vitamin E is a cofactor of selenium. Vitamin A and D work together, vitamin C and magnesium work in harmony. Also note that vitamin C is the king of the antioxidants. In 1932, Vitamin C was first isolated by C. G. King and W.A. Wagh from the University of Pittsburgh. The name ascorbic means "without scurvy" acid. The proper intake of vitamins and other important nutrients will contribute greatly to the achievement significant control of cancer and other deadly diseases. Vitamin C has extremely low toxicity and is remarkably free from harmful side effects, even when taken in large doses.[82]

Originally prepared by Albert Szent-Gyorgyi. (was pure ascorbic acid), it was not until 1932 his substance was identified as vitamin C. Dr Linus Pauling is considered the father of vitamin C because of all his research on it. His book published in 1979 indicated that 5 Milligrams (mg) of vitamin C per day was enough to prevent scurvy in most people. Thus, the MDR (minimum daily requirement is

---

[81] 26 Ibid p 19

[82] 27 Vitamin C p 103

5mg). However, larger amounts are required for optimal health and the recommended amount is 45 mg per day.[83]

Vitamin C is a required cofactor for, the production of collagen, connective tissue, cartilage, bones, teeth, blood vessel walls and capillaries. It also increases the absorption of inorganic iron. The biochemical properties of vitamin C include its functions a co-substrate in hydroxylation's requiring molecular oxygen, as in the hydroxylation of proline and lysine in the formation of collagen, of dopamine to norepinephrine, and of tryptophan to 5-hydroxytryptophane. Vitamin C also affects the functions of white blood cells, macrophages, immune responses, wound healing, and allergic reactions. Ascorbic acid increases the absorption of elemental iron when consumed together.[84] Vitamin C is an antioxidant that is required for at least 300 metabolic functions in the body, including tissue growth and repair, adrenal gland function and healthy gums. It is an important immune system protein. It is needed for the metabolism of folic acid, tyrosine, and phenylalanine. It can combine with toxic substances, such as certain heavy metals and render them harmless. Vitamin C works synergistically with both vitamin E and beta-carotene. Long term users of vitamins C and E in combination seem to have higher cognitive abilities as they age, as reported by a 2003 study.[85]

Research shows that, when the blood contains high

---

[83] Cancer & vitamin C p 100

[84] Epigenetics p 391

[85] Nutritional Healing p 26

levels of vitamin C, that the vitamin C has the tendency to purge lead from the body.

Normal calcium and phosphorus absorption and metabolism are dependent on proper levels of vitamin D. Blood levels of these ions are influenced by gastrointestinal absorption, skeletal metabolism, and renal excretion. These processes are predominantly under the control of vitamin D, parathyroid hormone and thyrocalcitonin.

An understanding of the close relationship between vitamin D and the parathyroid hormone is of major importance when considering the variety of effects enacted by vitamin D resulting from a variety of dietary and hormone conditions. Vitamin D is required for the full range of parathyroid hormone functions. Most consider vitamin D a hormone, as there are similarities between vitamin D and hormone activity. This vitamin has two active forms; D2 the plant source and D3 the animal source, which is created by exposing the skin deposits of cholesterol to UV sunlight.[86] Vitamin D has properties of both a vitamin and a hormone. It is especially important for the normal growth and development of bones and teeth in children. Current data indicates that vitamin D2 is as effective in maintaining vitamin D levels in the blood as vitamin D3. Vitamin D2 from food sources, requires conversion by the liver and then by the kidneys before becoming active.

Vitamin E was isolated in 1922. Vitamin E is a group of compounds referred to as alpha-tocopherols. It has antioxidant functions and protects cell membranes from

[86] Epigenetics p 377-378

oxidative-inflammation. It protects red blood cells from lysis, and in combination with the trace mineral selenium it can reduce the risk of certain cancers. It slows down the aging process, preserves the length of telomeres, reduces the risk of Alzheimer's disease, hypertrophic cardiomyopathy, muscular dystrophy, and cystic fibrosis.

Compared to the other fat-soluble vitamins, vitamin E is extremely safe when taken orally. Most adults have no problems ingesting up to 800 mg daily, showing no clinical or biochemical signs of toxicity.[87]

Vitamin E is a family of eight antioxidant compounds. These consist of four tocopherols, (alpha, beta, gamma, and delta), and four tocotrienols. The alpha-tocopherol form is the one found in the largest amounts in human blood. Small amounts of gamma-tocopherol are also present in the blood Alpha-tocopherol acts as an antioxidant in the human body, by inhibiting the oxidation of lipids and the formation of free radicals. Vitamin E is essential for life. It is difficult to get this nutrient from food alone. Supplements are recommended.[88]The d-alpha tocopherol form of vitamin E is the most potent and natural form. It is the one most recommended.

Vitamin K, isolated in 1939, is a fat-soluble vitamin that is required by the liver for the production of prothrombin and at least five other proteins and other biologically active substances essential for proper blood clotting and for the proper blood deposition of calcium in

---

[87] Ibid p 380

[88] Nutritional Healing p 28

the bones. About 50% of the required amount of vitamin K is produced by probiotic bacteria in the intestines.[89]

Vitamin K may help prevent osteoporosis. Additionally, it may protect the vascular system by preventing calcification in the arteries. The liver is a very efficient extractor of vitamin K. Vitamin K plays an important role in the intestines, and aids in converting glucose into glycogen for storage in the liver which can promote healthy liver function. Vitamin K is found in three forms. First as K1 from plants, and this is the dietary form of vitamin K. Second as K2, made by the intestinal bacteria it is also found in butter, cow liver, chicken, egg yolks, fermented soybean products and some cheeses. The third form is K3 which is the synthetic man-made product. Herbs that can supply vitamin K1 include alfalfa, green tea, kelp, nettle, oat straw, and shepherd's purse.

But the majority, of the body's supply of this vitamin is synthesized by the friendly bacteria normally present in the intestines, which comes as a result, of consuming soluble fiber.

Bioflavonoids are a diverse group of carbon compounds that are biologically active. These were isolated in 1936. The name flavonoid was derived from the Latin word, flavus, which means, the color yellow, which is common in flavonoids. While not generally considered essential nutrients, there is more than enough evidence to support a claim for essentiality.

Studies suggest that flavonoids produce an anti-inflammatory effect by way of a modulation, or through

---

[89] Epigenetics p 380

their ability to inhibit reactive oxygen compounds. They are also thought to inhibit the pro-inflammatory activity of enzymes that produce free radicals.

For example, flavonoid containing grape seed extracts have demonstrated protective antioxidant properties against reactive oxygen species in the gastrointestinal tract.

Studies at the Pauling Institute suggest that antioxidant effects in human blood produced by bioflavonoids are due to an increase in the production of uric acid related to the excretion of flavonoids.

Flavonoids are plant pigments that are found in flowers as yellow, red, and blue petals. These colors are attractants for pollinating insects and animals indirectly associated with the annual sexual cycle of plants. In (in vivo studies) supplementation with flavonoids has demonstrated clinical effectiveness for reducing or eliminating allergies, vascular and pulmonary inflammation, increasing ORAC (oxygen radical absorbance capacity) activity, improving gut functions, and supporting the immune system in cancer patients.

Bioflavonoids are essential for the absorption of vitamin C, and the two should be taken together. The human body does not produce bioflavonoids, they must be supplemented in the diet. Bioflavonoids act synergistically with vitamin C to protect and preserve the structure of capillaries.

Amino Acids

Different research studies indicate slightly different amounts of necessary amino acids, but most agree on 28. We will discuss twelve amino acids, which are: Isoleucine, Leucine, Lysine, Phenylalanine, Threonine, Tryptophan, Methionine, Histidine, Arginine, Taurine, Tyrosine, and Valine.

In the scientific community when the term "essential" is used, for nutrients, it usually means that the nutrient must be consumed in our diet. Nonessential nutrients are normally produced by our body. 80% of these are produced by the liver.

Amino acids are the basic structural unit of proteins and are second only to water content to the human body. Amino acids are also essential to the function of neurotransmission, transport, and biosynthesis. Amino acids are the chemical units, or "building blocks" as they are more commonly called, that make up protein. They are also the end products of protein digestion, or hydrolysis. Amino acids contain about 16% nitrogen. This is chemically what distinguishes them from the other two basic nutrients, sugar and fatty acids, which do not contain nitrogen. Protein provides the structures for all living things. Every living organism, from large animals to tiny microbes, all are composed of protein. Proteins are a necessary part to every living cell in our body.

Next to water, protein makes up the greatest proportion of our body weight. The proteins that humans require are not obtained directly from the diet. Dietary protein is normally broken down into its constituent amino acids, which the body then uses to build the specific proteins it needs. It is the amino acids rather than proteins that are the ere essential nutrients.[90]

Essential Fatty Acids – EFA'S

Essential Fatty Acids; there are 2 or 3, Omega 3's, Omega 6's, (Omega 9's). The main EFA is the Omega 3. EFAs are fatty acids that all vertebrates, including humans, must consume daily because they cannot be synthesized by the body, and are required for numerous bodily functions. Two essential fatty acids are identified. However, they were originally listed as vitamins. In 1929, studies indicated that these two EFAs were more properly classified as fats, rather than vitamins.

The term EFAs refers to long-chain fatty acids that are required for normal biological functions. The EFA group does not include fats that are primarily used as fuel.

The two EFAs that are normally considered essential are alpha linolenic acid (LNA) or Omega 3 and linoleic acid (LA) or Omega 6. LNA and LA are essential for humans but, our bodies do not produce them, which means we must either get them from our food or supplements. Note that EFAs attract oxygen. [91]

Basically, EFAs are involved with producing energy in our body from food substances and moving that

---

[90] Ibid p 54

[91] Ibid p 400

energy throughout our systems. They convert oxygen, electron transport, and energy in the process of oxidation. Oxidation is the central and most important moment to moment living process in our bodies. It is the burning of food to produce the energy required for our life processes.

LA and LNA appear to hold oxygen in our cell membranes where the oxygen acts as a barrier to viruses, fungi, bacteria, and other foreign organisms that cannot thrive in its presence. EFAs, are like oxygen magnets that can pull oxygen into our body, much like a magnet pulls iron filings. LA and LNA are thought to be involved with membranes. At the cell end, they apparently help transport oxygen from our red blood cells through plasma and across cell membranes into our cell walls to precise locations in our mitochondria, which use them in oxidation reactions to produce energy.[92]

It has been observed in a study of 96 boys, who had behavior and learning problems, had below normal levels of Omega-3 and Omega-6 EFAs. In this same study it was noted that the more that the boys were subjected to any use of antibiotics, the greater the behavioral problems. [93]

From LNA two fatty acids are formed, EPA and DHA which are vital for brain and eye development. In, order to produce these fatty acids from LNA the body requires vitamin C, B3, B6, magnesium, zinc, and some enzymes.[94]

It is generally accepted that a ratio of 2:1 of Omega-3

---

[92] Fats that Heal p 47

[93] Naturopathy 21st cent p 259

[94] Gut and Psych Syndrome p 267

to Omega-6 for supplements is recommended. Cod liver oil appears to have this ratio occurring naturally.

According to a new report in the Journal of Alzheimer's Disease, the brain is 60% fat, and of that fat 12% is EFA's. Studies performed in Amsterdam NL, on May 19th, 2017 show increased blood flow in regions of the brain associated with memory and learning for people with higher Omega-3 levels.

Enzymes

The late Dr Edward Howell, a physician and pioneer in enzyme research, called enzymes the "sparks of life". These energized protein molecules play a necessary role in virtually all the biochemical activities that occure in the body. They are essential for digesting food, for stimulating the brain, for providing cellular energy and for repairing all of the tissues, organs, and cells. Life as we currently know it could not exist without the action of enzymes.

Primarily, enzymes are catalysts. They are substrates that accelerate and precipitate the hundreds of thousands of biochemical reactions in the body that control life's processes.

Were it not for the catalytic action of enzymes, most of these actions would take place far too slowly to sustain life. Each enzyme has a specific function in the body that no other enzyme can fulfill. The substance on which an enzyme acts is called a substrate.

Because there is a different enzyme for every substrate, the body must produce a great number of enzymes.[95]

Enzymes assist in practically all bodily functions. Digestive enzymes break down food particles for energy. The clinical reaction is called hydrolysis, and it involves the use of water to break the chemical bonds, in order, to turn food into energy. Respiratory enzymes aid in eliminating carbon dioxide from the lungs. Enzymes assist the kidneys, liver, lungs, colon and skin in removing wastes and toxins from the body. Enzymes, prompt the oxidation of glucose, which creates energy for the cells. Enzymes also protect the blood from dangerous waste materials by converting them into forms that are easily eliminated by the body.

Enzymes are often divided into two groups. The digestive enzymes and the metabolic enzymes. Digestive enzymes are secreted along the gastrointestinal tract and break down foods allowing the nutrients to be absorbed into the blood for use in various bodily functions.

Of the macro-nutrients, carbohydrates, proteins and fats, most people experience the most trouble digesting fat, followed by protein and carbohydrates.

Life is not possible without enzymes. It is enzymes that are responsible, for the vast Majority, of all the biochemical reactions that bring our foods to maturity or ripeness. These enzymes will also digest the food in which they are contained when conditions are right for that to happen.

Enzymes are energy and energy, is defined in high

[95] Nutritional Healing p 72

school physics textbooks as "capacity to do work". Enzymes have the energy to perform the biochemical and physiological reactions that occur in all living things. The other components of our food supply specifically the protein, carbohydrates, fats, vitamins, and minerals (co-enzymes), are only building blocks. We work to obtain "nutritional objectivity" and use it to prevent disease and maintain health. The focus should be more on the gut and digestion as the place to begin healing or to maintain health. Prevention is always the best. Referring back to the work of Dr Edward Howell, who stressed the good results obtained with raw foods and fasting,[96] it is widely accepted that raw foods contain all the necessary enzymes to digest them.

The science of nutrition is a relative new comer in the scientific family. Nutrition was recognized as a distinct discipline in 1934 with the organization of the American Institute of Nutrition. Nutrition had to await advances in chemistry before maturing into a distinct discipline. Nutrition depends on and contributes to biochemistry, microbiology, physiology, cellular biology, medicine, and food science. Nutrition has been defined as the science of nourishing the body properly or the analysis of the effect of food on the living organism.

In the 18th century Lavoisier, become known as the father of nutrition and he initiated the chemical-analytical era. His work with guinea pigs was the first investigation of the relationship between heat production and oxygen use in the body. Since 1955 and the understanding of

[96] Enzymes p xxiii

the intricacies of cell structure, and complex and vital role that nutrients play in the growth, development, and maintenance of the cell is accumulating rapidly. Nourishment of cells is essential for nourishment of tissues. These tissues send nourishment back to the organs and ultimately the whole body. [97]

The increase of degenerative diseases, that occur primarily in those affluent societies where enzyme deficient foods are chosen instead of raw fruits and vegetables, have greatly focused the public's attention on the inability of the medical community to improve nutritional health. This just proves that if we are to improve our health it is up to us, we must take matters into our own hands. We must change our lifestyle and eating habits. We make choices every day, all day long. It is up to us to continue to study and research what is best for us. There are of course general guidelines, however we must find out what is best for us. Every "body" is different, one size does not fit all. We must find what works best for us. We are all unique; God made us all just a little bit different. It is up to us to find that difference. If it is to be it is up to me. Health is wealth, and each one of us has, the ability, to be very wealthy. However, it may not be easy and it may require some sacrifice and effort on our part. Proper diet and proper actions and proper attitude are not always popular with your friends and you may be on this path all alone. You will, have to make, the choice to do what is best for your health and the health of your family.

---

[97] Ibid p 7

This is the time to step up and be a leader and show others the way to health and happiness.

When choosing an enzyme supplement, be sure to address your specific digestion needs. The three main categories are: amylase, protease, and lipase. Amylase is in the saliva and the pancreatic and intestinal juices. It breaks down carbohydrates. Amylase is secreted from the parotid glands. This activity begins in the mouth by chewing. The types of amylase enzymes are: lactase which breaks down milk sugar, maltase which breaks down malt sugar, and sucrase which breaks down cane and beet sugars (sucrose). Protease is found in stomach juices and in the pancreatic and intestinal juices, and aids in the digestion of protein. Protease is secreted from the submandibular glands. Lipase is found in the stomach and pancreatic juices and aids in fat digestion.

Hydrochloric acid (stomach acid) is also important to proper digestion, as it works with the digestive enzymes as they perform their functions. Many people gain benefit from consuming raw foods, in addition to supplementing. Raw foods tend to aid the body in maintaining its enzyme reserves. Enzymes are made from protein; it is therefore essential to consume adequate amounts of protein in your diet.

It is generally accepted that a balanced diet should consist of roughly 50% carbohydrate, 30% protein, and 20% fat [98].

---

[98] Enzymes p xxi

Hormones / Endocrine System

Hormones are the messenger molecules of the endocrine system, including the fat cells. The endocrine system refers to the collection of glands in the body that secrete hormones directly into the blood stream to be transported toward distant target organs.

The major endocrine glands include the pineal gland, the pituitary gland and hypothalamus, the thyroid and parathyroid glands, the thymus gland, the pancreas, the ovaries and testes, and the adrenal gland. The endocrine system has five major functions; First is to regulate the metabolism. Second to maintain, salt, water, and nutrient balance. Third to control responses to stress. Forth to regulate growth development and reproduction. Fifth to produce more hormones. It is interesting to note the correlation of the chakras in the ancient Ayurvedic Medicine of India, to our modern-day endocrine system. The seven chakras line up with the seven main glandular locations of the endocrine system. Hormones are normal constituents of the human body, which could be described as chemical messengers and that they function in the control of the metabolic process.[99] Environment chemicals and heavy metals are well known hormonal disruptors. Toxins can affect many of the major weight control hormones besides the thyroid. Estrogens, testosterone, cortisol, insulin, growth hormone and leptin. Toxins also interfere with the stress response, our automatic nervous system (ANS) and can alter the normal circadian rhythms

---

[99] Cancer & vitamin C p 70

that control our eating behavior and our sleep patterns. Toxins are certainly not the only factor in our obesity epidemic, but they must be considered. It is important that we learn how to minimize our exposure to environmental toxins and to use food, exercise, and supplements, to aid in controlling the environment.[100]

Hormones speed up or slow down chemical reactions necessary for normalizing conditions in our body. Note that, infectious agents stimulate the production of pituitary and thyroid hormones that increase oxidation and metabolic rate, that speed up chemical reactions. Interesting that, all fats act like hormones.[101] Fat cells are known as adipocytes. Your, fat cells are endocrine cells that produce many hormones. Most of the biology of fat cells are controlled by the quality and type of food you eat. Of all the endocrine disorders, diabetes is the most common, it is also a disease in which nutrition plays a major role in treatment. Working closely with the nervous system the endocrine system controls such diverse activities, as reproduction, metabolism, growth, sleep cycles, maintenance of the body's salt, water, chemical balance and response to stress. In addition to the endocrine system, several other organs, especially the brain, kidneys, lungs, heart, and intestinal lining also secrete hormones. Hormones produced by the thyroid gland control, the body's metabolic rate. Thyroid hormones stimulate growth and are essential of normal development of the

---

[100] Ultra -metabolism p 384

[101] Fats that Heal p 370

nervous system. They also speed up the action of insulin and enhance the body's response to catecholamines.[102]

Methodology, I am seeking more qualitive than quantitative research for my dissertation. Topic proposal 3%, prior current research 55%, writing editing and proofing 42%

[102] Total Nutrition pp 496 & 514

Chapter 4

Lifestyle, Obesity, Illness, Diet, Gut Health

If we are to avoid the suffering of arthritis, *Obesity*, premenstrual syndrome, and certain types of mental illness, we need take this last and least well understood area of human nutrition into full account in our food choices. To do so we must, be more educated and become more informed.

In chapter one we covered sugar and its ill effects. It inhibits the functions of the immune system. Sugars lack the vitamins and minerals required for their own metabolism and must draw on our body's stores of these nutrients. Once depleted, the body becomes less able to carry out other functions. As a result, cholesterol levels rise, our metabolic rate decreases, fats burn more slowly, we feel less like exercising, and we may become *Obese.*

Obesity increases the risk of diabetes, cardiovascular disease, and cancer. People suffering from degenerative diseases such as *Obesity*, cancer, cardiovascular disease, and liver degeneration normally have lower EFA's stored in their tissues.

The edge, of *Obesity*, is considered, to be 23% of body weight as fat for men and 32% of body weight as fat for women.[103] Researchers who favor the TG theory, state that, whatever increases the TG will increase our risk of CVD. Besides sugar, fats, excess calories, and Obesity, lack of exercise increases TG levels, because excess calories that turn to fat are not worked off.

About 30% of all affluent people are overweight and 10% are obese. Both are forms of fatty degeneration. Omega 3 fatty acids will increase the metabolic rate, and aid in weight control. There is some evidence that Obesity is the result of gross overeating in only about 10% of cases. The other 90% are lacking exercise and choosing foods lacking important essential nutrients. Omega 3's will increase energy levels and make you feel more alive, so that you will feel like exercising and that will aid you with your weight control.

America spends more on health care annually, than all the other nations combined. As of 2013 the only health category that the United States is ranked number one in is *Obesity. Obesity* is universally believed to have a genetic cause and that it can only be dealt with by means of gastric by-pass surgery, exercise and vigilant calorie and portion control. These, beliefs are mostly untrue.

In 1980, 15% of American adults were *Obese*. In 2000, the obesity rate increased upward and more than doubled to 30.5%. By 2012 the rate had crept up to 40%.

---

[103] Fats that Heal pp 123 & 156

It should be obvious to the average observer that the medical theory that *Obesity* is due to "eating too much and lack of exercise" along with the practice of *Obesity* control with drugs and surgery had failed. *Obesity* is not a genetically-generated disease, not just a result of lack of exercise, and not a disease that can be attributed solely to overeating. *Obesity* is caused by a deficiency of minerals that universally manifests symptomatically as cravings, binge eating and munchies.[104]

Artificial sweeteners alter gut flora or bacteria to promote *Obesity* and type 2 diabetes. In 1942 Dr Frederick Stare founded the Department of Nutrition at the Harvard School of Public Health and served as its chairman until retirement in 1976. He helped to entrench the idea that *Obesity* was nothing more than a matter of energy imbalance. In his books, column, and writings he encouraged the public to accept the fact that there are no fattening foods, or sliming foods, we are just eating too much food.

In 1980, the U.S. Dietary Guidelines have warned us and against the dangers of eating fat and implored us to eat less fat. Not only, is this advice not working it is causing us harm. It turns out that by eating less fat, results in more *Obesity* and disease. *Obesity* continues its rise and it is projected that one out every two Americans will be *Obese* by 2050. That equals to 50% of our population. Sugar sweetened drinks have been proven to cause *Obesity*, heart disease, type 2 diabetes and cancer.[105]

---

[104] Epigenetics p 359

[105] Eat Fat Get Thin p 32 d

It is now evident from the research, that sugars and refined carbs are the true causes of *Obesity* and heart disease not fats. In 2015, US Dietary Guidelines Advisory Committee, after reviewing all the research, this group of scientists failed to find any reason to limit total fat or cholesterol in the diet. Low fat diets have unintended consequences, turning people away from healthy high fat foods and towards foods with added sugars and refined grains.

Early in the twentieth century, German diabetes specialist, Carl von Noorden, believed in the 'metabolic hypothesis' but changed his position. In his early work, he postulated that *Obesity* was a pre-diabetic state and said that "*Obese* individuals of this type have already altered their metabolism for sugar, but instead of excreting the sugar in the urine, they transfer it to the fat-producing parts of the body."

In 1953, in a report in the New England Journal of Medicine, titled "A reorientation on *Obesity*", Dr Alfred Pennington argued that *Obesity* was caused by the hormonal effects of carbs and could be treated by restricting carbs, without worrying about fat and protein.

In 1977, a study published in the American Journal of Clinical Nutrition showed that the composition, of your diet, could have profoundly different effects on human biology even though the calories consumed were identical.

In 2002, Dr Walter Willett, of the Harvard School of Public Health, summarized all the research on fat and Obesity and found no connection. He stated "that diets high in fat do not appear to be the primary of the high prevalence of excess body fat in our society and reductions in fat will not be the solution."

Much of the modern world consumes a diet rich in processed grains, oils, sweets and animal products. In the US less than five percent of total calories consumed are from fresh fruits, vegetables, seeds and nuts. All the foods that contain the richest amounts of micronutrients.

The standard American diet (SAD) contains an overabundance of calories, but a very low nutrient per-calorie intake, making the US a chronically overfed but malnourished nation.

Glycemic Index and the Glycemic Load

Natural healthy carbs contain more micronutrients and are loaded with fiber and so called resistant starch, both of which keep these foods' glycemic index and calorie density low.

The glycemic index (GI) is an index of their rate of conversion to sugar and calories in the blood after eating a carbohydrate-containing food, the higher the number, the greater the blood response. A low glycemic index diet is healthier.[106]

The glycemic load (GL). Most low GL foods are also high in phytonutrients. GL measures the real response of our blood sugar and your insulin level to an entire meal. The GL is the effect a total meal has on your blood sugar and in not only related to the original form of the

[106] Nutritional Healing p 6

carbohydrate. The glycemic load is the best measure of how quickly or slowly you absorb the sugars in all the foods you eat in any given meal. The GL also takes, into account, all the factors including the effect that mixing carbs, fats, protein and fiber has on your metabolism. GL meals contain a combination of foods that either do not have many carbs to begin with or whose carbs are absorbed slowly and do not lead to rapid rising and high blood sugar levels that promote obesity an aging.[107]

Natural healthy carbs contain resistant starch, it is like fiber, in that it is resistant to digestion and does not break down to glucose or simple sugars. It is classified as a prebiotic which promotes the growth of beneficial bacteria, or probiotics, in the digestive tract. These bacteria then break down the resistant starch into favorable compounds that improve our immune system function and reduce cancer risk.[108]

The healthiest starches are in high-fiber, natural foods. Generally, good carbs are complex carbohydrates and bad carbs are refined or simple carbohydrates. For example, whole wheat products have a lower glycemic effect than white flour products and greater amounts of trace minerals.

---

[107] Ultra-Metabolism p 95

[108] Super Immunity p 111

Gut Health and The Immune System

Good gut health equals good health. Good gut health gives us a strong immune system. Without, a strong immune system we are susceptible to any and every virus that enters our environment. All diseases begin in the gut. Many of our digestive problems begin at weaning time when the breast or mothers milk gets replaced with formula milk and other foods get introduced. Important to note that constipated individuals with large amounts of old compacted feces which stay in the in the digestive tract, provide a fertile rotting environment for all sorts of parasites, fungi, bacteria, and viruses to breed and thrive which constantly produces many toxic substances which are absorbed into the bloodstream.

The last portion of the small intestine is, the ileum, not much digestion happens in the ileum. The walls of the ilium are packed with large numbers of lymph nodes. Lymph nodes are a very important part of our immune system. They preform two important functions; first they filter the lymph and remove bacteria, viruses, fungi dead cells (including cancer cells) and various toxins, the second function is production of fighting infections. Lymph nodes are made primarily of lymphocytes.

Several studies and clinical experiences have produced a substantial amount of research linking schizophrenia with digestive abnormalities. Many of these cases can be traced to early childhood.[109]

The human body is like a planet inhabited by large

[109] Gut & Psychology Syndrome p 11-13

numbers of various micro-creatures. The diversity and richness of this life on every one of us is probably as amazing as the life on earth itself. Our digestive system, skin, eyes, respiratory and excretory organs are happily coexisting with trillions of invisible lodgers, making one ecosystem of macro—and micro-life, living together in harmony. It is a symbiotic relationship, where neither party can live without the other. [110]

As it turns out, gut thoughts and feelings are not a fanciful notion but a physiological fact. Rather than one brain found in our head, scientists have determined that we have two brains., and the second, being located, in the digestive tract. The gut brain is known as the enteric nervous system (ENS) and is housed under the mucosal lining and between the muscular layers of the esophagus, the stomach, and the small and large intestines.

When scientists counted the number of nerve cells in the gut brain, they found it contained over one hundred million neurons, which is more than the number, of nerve cells in the spinal cord. Researchers have observed a significantly greater flow of neural traffic from the ENS to the head-brain than from the head-brain to the ENS. This indicates that rather than head informing the digestive system what to eat and how to metabolize, the locus is stationed in our belly.

The entire digestive tract is lined with cells that produce and receive a variety of neuropeptides and neurochemicals, the same substances that were thought to be found in the brain alone. The list includes, serotonin,

---

[110] Ibid p 15

dopamine, norepinephrine, and glutamate. There are also many hormones and chemicals in the gut previously thought to exist only in the gut are also active in the brain, including insulin.

In Japan the mid section is considered the seat of wisdom and the locus of our center of gravity, both physical and spiritual. Known as the hara, it is centered around a point just below the navel. The first practical step necessary tor tapping into ENS intelligence is to breath into the belly. The saying, 'where attention goes energy flows', certainly applies here. Breathing into the belly increases its oxygen uptake and activates the ENS. There are several old traditions which use belly breathing, such as yoga and taichi. Flood the belly with oxygen and the gut brain is more responsive and alert.[111]

The largest colonies of microbes live in our digestive system. The number of functions they fulfill in our bodies are so vital to us that if our gut got sterilized, we would probably not survive. In a healthy body the microbial world is fairly stable and adaptable to changes in their environment.

Gut micro-flora can be divided into three groups; the essential or beneficial flora, this is the most important group and the most numerous in a healthy individual. These bacteria are often referred to as our in indigenous friendly bacteria. They include Peptostreptococci, and Enterococci.[112]

---

[111] The Slow Down Diet pp 71,72,75

[112] Gut and Psychology p 15

It appears that the most commonly used probiotics are Lactobacillis Acidophilus and Bifidobacteria Bifidum.[113]

The second group is the opportunistic flora, which are a large group of various microbes. There are around 500 various species of microbes known to science so far which can be found in the human gut. Ina healthy person their numbers are normally limited and are tightly controlled by the beneficial flora.

The third group of gut micro-flora is the transitional flora, and these include various microbes which we swallow daily with our food and drink. They are usually non-fermenting gram-negative bacilli from the environment. When the gut is well protected by beneficial bacteria, this group of microbes through our digestive tract without doing any harm.

Our healthy indigenous gut flora has a good ability to neutralize nitrates, indoles, phenols, skatol, ksenobiotics and many other toxic substances. Gut flora has the ability to inactivate histamine and chelate heavy metals and other poisons including one common ubiquitous fungus candida albicans. Gut flora is the housekeeper of the digestive system. Without healthy gut flora we can not digest certain foods such as dairy products, which have casomorphins and wheat products which contain gluteomorphins, which are, morphine like substances.[114]

Casomorphins, which is the protein in dairy, contain opiates which have a calming effect and cause the brain

---

[113] Ultra-Prevention p 270

[114] Gut and Psychology p 21

to release dopamine leading to a sense of reward and pleasure.[115]

Probiotics are found in many fermented foods. By restoring the normal gut flora, they reduce overall immune activation, and have been proven effective in many inflammatory diseases. They work by balancing the gut associated lymphoid tissue (GALT).[116]

The main purpose of a digestive system is to be able to digest and absorb food. Scientific and clinical experience shows that without healthy gut flora the digestive system cannot fulfil these functions efficiently.

By-products of bacterial activity in the gut are very important in transporting minerals, vitamins, water, gases and many other nutrients through the gut wall into the blood stream.

Certain ingredients in food cannot be digested by a human gut at all without the help of beneficial bacteria. The protein in dairy is casein, the protein in wheat is gluten. The digestive juices in the stomach split these proteins into peptides. These peptides then move into the small intestine where the next stage of their digestion happens. They get subjected to the pancreatic juices, upon reaching the intestinal wall they are broken down by enzymes, called peptidases. Normal digestion and absorption of food is probably impossible without well balanced gut flora.[117]

---

[115] The Cheese Trap p 49

[116] Ultra-metabolism p 603

[117] Gut and Psychology p 21

Our Immune System

Important to note that 60% to 80% of our immune functions are in the gut, the intestines. Many experts believe that toxic elements from the environment can increase the body's exposure to free radicals. Damage from free radicals is an important factor in causing the uncontrolled cellular growth that is characteristic of cancer. It is believed that this is the result of and impaired immune system.[118]

The health-promoting bacteria, called probiotics are normal inhabitants of the human gastrointestinal tract. Amazingly, bowel bacteria cells comprise approximately ninety five percent of the total number cells in the human body. These bowel inhabitants play a critical role in the health of our immune system.[119]

Good intestinal bacteria secrets antibacterial substances that prevent disease causing bacteria from taking hold in your body. The health-promoting bacteria crowed out and prevent the development of pathogens looking to take hold as bacterial illnesses.

The epithelial surfaces of the digestive system inhabited by large numbers of bacteria can truly be described as the cradle of the immune system both systematic and mucosal. Essential or beneficial bacteria in our digestive system engage a very important member of the immune system, the lymphoid tissue of the gut wall, and take part in the production of huge numbers of lymphocytes and

---

[118] Nutritional Healing p 282

[119] Super Immunity p 89

immunoglobins. In the cell wall of Bifido-bacteria, there is a substance called "Muramil Dipetide" which activates synthesis of one of the most important groups of immune cells, the lymphocytes. As a result, a healthy gut wall is literally infiltrated, with lymphocytes, ready to protect the body from any invader.[120]

Lymphocytes in the gut wall produce immunoglobulin A (IgA). Secretory IgA is a substance which is produced by lymphocytes in all mucous membranes in the body and excreted in body fluids. Its job is to protect mucous membranes by destroying and inactivating invading bacteria, viruses, fungi and parasites. It is one of the immune system's ways of dealing with the unwelcome invaders coming with food and drink into our digestive system.[121]

The IgA levels effect on the immune system, seem to increase with positive thoughts and responses, using positive affirmations can aid the system in producing a positive attitude.[122]

Food, Diet, Nutrition

Earlier we spoke of free radicals. Free radicals are controlled by antioxidants. We will provide a scale that will assist in the understanding of the Oxygen Radical Absorbance Capacity (ORAC).

---

[120] Gut and Psychology p 27

[121] Ibid p 28

[122] Change your brain, change your life p 156

So many Americans are constantly on the go, consuming a lot of junk in the form of processed foods and fast foods. Many people today just do not know how to choose and cook organic, antioxidant rich fruits and vegetables. Most people want to make healthy decisions but just need some guidance and education in making that happen. The intention of this research paper remains the same. The focus is to keep healthy eating as simple as by clear nutritional information for the average person to be able to use on a, daily basis.

It is important to know that the ORAC value is a method of measuring the antioxidant capacity of various foods. The scale is primarily used to determine which foods have the highest antioxidant values. As a society we are constantly receiving conflicting nutritional advise from a variety of sources, the medical community our doctors and health practitioners, the news media, diet books, the pharmaceutical industry, government initiatives and the industrialized corporate food giants.[123]

Our society is dominated by the Western Diet (SAD), the Western food pattern it which consists of a lot of processed foods, meats, fats, and sugar. Additionally, it contains preservatives, additives, hormones, and genetically modified ingredients (GMOS) that the human body does not recognize as food and is unable to metabolize properly, if at all.

Research shows that people who follow this diet (SAD), commonly end up with chronic diseases; obesity, heart disease, diabetes, and cancer. Of the top ten health

---

[123] The antioxidant counter pp 9-12

related killers in the Western population, five are chronic diseases directly related to the diet.[124] Our current diet consists of corn and soy, which make up more than one third and as much as two thirds of our daily diet, include sugar in this diet and carbohydrates make up fifty to seventy percent of the daily American Diet. Then the balance of the diet consists of processed meats, fats and some fruits and vegetables.

Since the younger generations, have been consuming this SAD diet since childhood, we find ourselves in a time when children will die of, chronic diseases. before their parents.[125] Children are still gaining weight and developing diabetes prior to age 15.

I believe that people are more likely to do something when they understand why it is important. The two prior statements should stress the importance of why it is necessary to change our diets, sooner rather than later.

The formation of free radicals in the body is a normal aging process, happening as a result of breathing. We can slow down the process with the aid of antioxidants. Three of these most powerful antioxidants available to us are vitamin C, vitamin E and the mineral selenium. Current research has brought another element to our attention, this is glutathione. Glutathione has been determined to be the second most powerful antioxidant available next to vitamin C. It is important to restate the fact that none of the vitamins, amino acids, enzymes or minerals work alone they all work together.

---

[124] Ibid p 14

[125] Ibid p 17

The recommended human consumption is about 5000 ORAC units daily.[126] The higher the better. A unit is approximately 100gr or about 3.75oz, which is just under ½ cup.

(4oz = ½ cup). Eat eight to ten servings of colorful fruits and vegetables every day.

[126] Eat Right Now p 101

Chapter 5

Big Food, Big Pharma, Big Chemical Companies (Monsanto), GMO's, High fructose corn syrup, Exercise, Herbs, Sleep, WaterBelief Systems and Stress

Conclusions
Big Food

All, of the processed food products that are being sold to the American people are mostly junk. Antinutrients, empty calories loaded with chemicals and pesticides.

We need to and optimize our metabolism no matter what we choose to eat. We need to slow down and allow our body to bring the gifts of the soul into our dietary world and awaken the inner fire that is the true source of our power. We refer, to our midsection, the center of metabolic activity in the body as the solar plexus, which in Latin means “a gathering place for the sun”.

The dizzying pace at which our culture propels its self is contrary to a healthy and happy life. We are moving at a speed that has us move unconsciously through our day, that pushes us beyond the body’s natural capacity and leaves us unfulfilled and exhausted by day’s end. It results in meals eaten under physiologic stress-response, which diminishes, our calorie-burning powers. It shortens our breathing, which results in less oxygen intake and more fat accumulation.

We are consuming digestive aids and pain killers that yield debilitating side effects. We avail ourselves to medical therapies that never truly address the reasons for our bodily breakdowns. We punish ourselves with excessive exercise for the crime of eating.

The remedy for, this speed disease, is just slowing down. Our frame of mind directly impacts metabolism to such a degree that what we think and feel profoundly influences how we digest our food.[127]

Big Pharma

It is time to move away from fake food and fast food and return to whole organic food. Drug companies would have you believe that ‘bad’ of LDL low cholesterol, is the single biggest factor in the development of heart disease.

---

[127] The Slow Down Diet p 3

But the truth is that the determining factor is the ratio of your total cholesterol to your good or HDL cholesterol.

The pharmaceutical industry promotes this unfounded belief, not because there is scientific evidence, but because the main class of drugs available for treating high cholesterol are statins, which mostly lower LDL. These drugs are among the biggest selling drugs in history. All this equals profits.[128]

Big Government

On line article by Kristin Leutwyler, 1/10/01 – New study from the Union for Concerned Scientists, meat producers feed some 25 million pounds of antibiotics to chickens, pigs and cows for non-therapeutic purposes each year.

On line article by Maryn McKenna, 12/08/10 – FDA estimates US livestock get 29 million pounds of antibiotics per year. This was for 2009. It is estimated that 24.6 million pounds per year are used only for 'non-therapeutic purposes", that is to make animals grow to market weight faster and to prevent them catching diseases in close quarters of confinement agriculture.

Science and Innovation on line article by Maryn McKenna, 4/15/15 – FDA is again attempting to control how much growth promoters are used in US. Stats from 2013 show that 32.6 million pounds of antibiotics sold in the US for use in animals. Do the math, that is an

[128] i2 Ultra-Metabolism p 68

increase of 7.6 million pounds of antibiotics used from 2001 thru 2013.

If agriculture continues to use the drugs that are functionally identical to drugs that are important in human medicine, the bacteria will become resistant to those drugs when they are used in animals and this will also affect humans, making common diseases difficult or impossible to treat.

In 1992 president Clinton allowed the Food and Drug Administration (FDA) to accept millions of dollars in drug company money to help approve drugs, a practice that continues today.[129]

The United States Department of Agriculture (USDA), and the health 312 million people of the United States, which the USDA is charged with safeguarding is on one side and on the other side are the three hundred or so companies that form the $1 trillion industry of food manufacturing, companies that the USDA feels obligate to placate and nurture.[130] Which means that the American people lose out again.

Big Chemical Companies

On line article by Emily Elert, Environmental Health News 4/22/10 – As of the date of this article it was thirty-eight years since DDT was banned, Americans still consume traces of the chemical. Residues are ubiquitous

---

[129] Fight for your Health p 1

[130] Salt Sugar Fat p 213

in US food, particularly in dairy products meat and fish. That would bring the time frame current to forty-six years. DDT was originally banned 1972.

According to Dr Mark Hyman MD, DDT is in 100% of our beef 93% of processed cheese, hot dogs, bologna, turkey, and ice cream.[131]

Glyphosate resistant crops were introduced by *Monsanto.* These crops are known as *Roundup ready.* In 2015 it was estimated that *Roundup* was used on 80% of the crops worldwide thus we have GMO crops. Additional research indicates that *Roundup* kill off microbes in the soil helping to make it sterile.[132] Much of the corn, soy, cotton, sugar beets, canola and alfalfa grown in the US are drenched with *Roundup.* A study from MIT showed glyphosate interferes with human digestion, biosynthesis of nutrients, resulting in obesity, diabetes, heart, depression, autism, infertility, cancer and Alzheimer's disease.[133]

Herbs

Herbs are Medicine, when an herbal remedy is correctly matched to the patient, the result can be dramatic. Each plant has its own unique quality. Herbalism is a living art, continually evolving, incorporating new discoveries and

[131] Ultra-Metabolism p 377
[132] Buy and Eat Organic
[133] Clean up your Diet

assimilating older, less familiar traditions as populations shift and cultures collide.[134]

Hippocrates (circa 400 B.C.), who studied in Egypt. He stated, "That only nature could cure, and the province of the Physician was merely to assist, to make the healing more pleasant or less painful". He also said, "First do no harm".[135] He believed in treating the whole person. He used thirty-nine herbs along with diet, fasting and exercise. He believed that disease came from a disharmony with nature. He believed in sound mind and body.

Chakara, founder the of the Aruvedic Medical system and the oldest of Naturopathic Physician known, advocated air and water therapies.

The Romans emphasized more herbs, they employed 150 herbs mostly in inhalation therapy by burning the herbs. The Roman, Pliny believed that for every disease there had to be a plant to cure it. The Romans had three doctors, "Dr Quiet, Dr Diet, and Dr Happiness".

Exercise

Exercise in moderation, strenuous exercise may contribute to stress. Walking for fifteen to twenty minutes daily is excellent for your body, mind, spirit and heart. The body is made for movement, our current culture sits too much, and this contributes to many of our chronic health problems. The lymph system only works, when

---

[134] Radical healing p 21

[135] Naturopathy in the 21st Cent. P 1

we move, our bodies. Our lymph system removes toxins from our bodies.

According to Mayo Clinic and Harvard School of Medicine, Sitting is the new smoking. Resistance training is also necessary to get oxygen into our muscles. Use or lose it, if you do not use your muscles they will get weak. Exercise will give you more energy to do the things you like to do, and you will feel good doing them.

Foods to avoid, GMO's

In an ideal situation we will avoid all processed, prepackaged fake foods. A great place to start is with sugar. Specifically, High fructose corn syrup if the package reads corn syrup it is the same thing. Corn syrup is used by the soft drink industry. Avoid soft drinks and drink more water.

Drink eight to ten glasses of water a day. Our body is mostly water. Our brain is 80% water, healthy brain, healthy body.[136] A good formula is to take your body weight in pounds and divide by 50%, then consume that amount in ounces daily.

Avoid "partially hydrogenated" oils which turn into trans-fatty acids in your body.[137]

GMO's are genetically modified organisms, aka food, 80% of the corn, soy and sugar beets are GMO's. Our bodies do not always recognize these altered products as food.

---

[136] Change your Brain p 377

[137] Ultra-Metabolism p 80

Conclusion

One of the greatest contributors to our ill health is stress, it is difficult to relax if you have not had a, good nights, sleep. It is recommended that a healthy person should get between six to eight hours sleep each night. There are five stages of sleep; stage one, drifting off to sleep, our brain waves are moving from beta, to alpha, to theta. Stage two body temperature drops, breath and heart rate slow down. Stage three deeper and slower brain waves take over moving from light to deep sleep. Stage four is the deepest stage of sleep, your brain goes into theta and your body systems can be devoted to healing and repair. Stage five is the Rapid Eye Movement stage of sleep (REM) this is when dreaming occurs. After the first cycle of stage one thru five. The body will repeat stages two thru five. This will occur every ninety minutes.[138]

We need to Slow down and breath when we are eating. Taste our food when we eat we should only eat, no TV, no reading, just sit down and enjoy our food. There is no good or bad food only fake food. (including fast food).

[138] On line article by Dr Xu

It is called fast food because it speeds you up to an early grave.

When you eat try to eat a rainbow of colors on your plate, the more color the better. Slow down and breath, learn to move from stress to relaxation. There are several ways to accomplish this, meditation, yoga, tai chi, Qi gong,

We cannot change others, however, we can ourselves.

This is the time to change our thinking, our attitudes, our diets, and our lifestyles. This is the first day of the rest of our lives so do something, anything to make your life better, take small steps. Start with one thing today, make a habit and add something else later. We are victims of habit, so all that is necessary is to create some new habits. It is commonly accepted that it takes twenty-one to twenty-eight days to create a new habit.

Health is everyone's most, valuable asset and you are the CEO of these assets. Health is our birthright, your first freedom.

It is not the responsibility of the government, the medical community, or the pharmaceutical industry, we are responsible for our own health and wellbeing. If we change our lifestyle, we change. We can change our buying habits, vote with our dollars. Buy local, buy organic, Buy American. Stress is a major problem, especially stress related to debt. Work to become debt free and eliminate much of your stress. Lifestyle changes include asking the question, do I need this, or do I just want this.

Lifestyle changes include, proper diet, proper attitude, prayer, meditation, gratitude, exercise, and generosity. Gandhi said, "to be the change you seek". Regarding health, Ben Franklin said "An ounce of prevention is worth a pound of cure".

Using the research and historical data cited, I have endeavored to present to the American people with the information that is necessary for us to change and that we have the ability, to change for the better. We have the power. Napoleon Hill said "That whatever the mind can conceive and believe it can achieve"

The magic word in that sentence is believe. Our belief system is paramount. Bruce Lipton PhD states that "Your biology is not your destiny"

The mind is the most powerful pharmacy.

Thinking makes it so. Whatever you think about you bring about. Maintain a positive mental attitude. If you think you can, or you think you cannot, you are right.

Give yourself the best nutrition you can, and let the healing be done by the greatest physician of all... Your own body.

God Bless America

Footnotes

Introduction

1. Hyman, Mark – Ultra-metabolism p 17, 2006
2. Mayo Clinic
3. Moss, Michael – Salt, Sugar, Fat p xxvii, 2014
4. Hill, Napoleon – Think and Grow Rich p 32, 1963
5. Amen, Daniel G – Change your Brain, Change your Life p 61, 1998
6. Maurer, Robert – The Kaizen Way p 47, 2014

Chapter 1

7. Duffy, William – Sugar Blues p 46, 1975
8. Hyman, Mark – Ultra- metabolism p 409, 2006
9. Moss, Michael – Salt, Sugar, Fat p 36, 2014
10. Ibid – p 37
11. Erasmus, Udo – Fats that Heal, Fats that Kill p 123, 1993
12. Ibid – p 37
13. Duffy, William – Sugar Blues p 22, 1975
14. Moss, Michael – Salt, Sugar, Fat p 4, 2014
15. Duffy, William – Sugar Blues p 181, 1975
16. Erasmus, Udo – Fats that Heal, Fats that Kill p 333, 1993
17. Duffy, William – Sugar Blues p147,1975

18. Ibid, pp 137-138
19. Erasmus, Udo – Fats that Heal, Fats that Kill p 243, 1993
20. Ibid – p318
21. Petrucciani, Lori & White, Bea – Bring Back Vitality pp 72-73, 2013
22. Duffy, William – Sugar Blues p 147, 1975
23. Ibid – p 179
24. Ibid – pp 198, 199
25. Ibid – p 142
26. Ballentine, Rudolph – Diet & Nutrition p 20, 2007
27. Wallach, Joel – Epigenetics p 118, 2014
28. Ballentine, Rudolph – Diet & Nutrition p 76, 2007
29. Erasmus, Udo – Fats that Heal, Fats that Kill p 76, 1993
30. Ballentine, Rudolph- Diet and Nutrition p 75
31. Ibid – p 33
32. Duffy William – Sugar Blues p 28
33. Ibid – p 30
34. Moss, Michael – Salt, Sugar, Fat p 130, 2014

Chapter 2

35. Lois N Manger - History of Medicine p 2, 1992
36. Ibid
37. Joel Fuhrman -Super Immunity pp 11-12, 2011
38. Lois n Manger – History of Medicine p 30
39. Joel Fuhrman – Super Immunity p 13
40. Lois N Manger – History of Medicine p 41
41. Ibid p 43
42. Ibid p 47
43. Ibid p 51
44. Ibid p 65
45. Ibid p 68
46. Ibid p 71
47. Ibid p 83
48. Ibid pp 86-90

49. Joel Wallach – Epigenetics p 120
50. Robert J Thiel – Naturopathy, 21st Century p 195, 2000
51. Ibid p 196
52. Rudolph Ballentine -Diet and Nutrition p 24
53. Robert J Thiel – Naturopathy 21st Century p 17
54. Rudolph Ballentine – Diet and Nutrition pp 33-36
55. Robert J Thiel – Naturopathy 21st Century p 195

Chapter 3

56. Joel Wallach – Epigenetics p 208
57. Ibid p 209
58. Ibid p 236
59. Ibid p 288
60. James L Chestnut – Innate Diet p 150-151, 2004
61. Joel Wallach – Epigenetics p 289
62. Ibid p 319
63. Ibid p 405
64. Ibid
65. Ibid p 403
66. Ibid p 404
67. Ibid p 402
68. Phyllis A Balch – Nutritional Healing p 32, 2010
69. Ibid p 33
70. Ibid p 40
71. Wikipedia- dictionary
72. Phyllis A Balch – Nutritional Healing p 23
73. Joel Wallach – Epigenetics p 387
74. Ibid p 388
75. Phyllis A Balch – Nutritional Healing p 24
76. Ibid
77. Joel Wallach – Epigenetics p 392
78. Phyllis A Balch – Nutritional Healing p 25
79. Joel Wallach - Epigenetics pp 373-374
80. Phyllis A Balch - Nutritional Healing p 18

81. Ibid p 19
82. Joel Wallach – Epigenetics p 390
83. Linus Pauling / Ewan Cameron – Cancer & vitamin C p 100, 1979
84. Joel Wallach – Epigenetics p 391
85. Phyllis A Balch – Nutritional Healing p 26
86. Joel Wallach – Epigenetics pp 377-378
87. Ibid p 380
88. Phyllis A Balch – Nutritional Healing p 28
89. Joel Wallach – Epigenetics p 380
90. Phyllis A Balch – Nutritional Healing p 54
91. Joel Wallach – Epigenetics p 400
92. Fats that Heal p 47
93. Robert J Thiel – Naturopathy 21st Cent. P 259
94. Natasha Campbell-McBride – Gut & Psych Syndrome p 267
95. Phyllis A Balch – Nutritional Healing p 72
96. Howard F Loomis – Enzymes p xxiii
97. Ibid p 7
98. Ibid p xxi
99. Linus Pauling – Cancer & vitamin C p 70
100. Mark Hyman – Ultra-Metabolism p 384
101. Udo Erasmus - Fats that Heal p 370
102. Victor Herbert – Total Nutrition pp 496 & 514, 1959

Chapter 4

103. Udo Erasmus – Fats that Heal pp 123 & 156
104. Joel Wallach - Epigenetics p 359
105. Mark Hyman – Eat Fat, get Thin p 32, 2016
106. Phyllis A Balch – Nutritional Healing p 6
107. Mark Hyman – Ultra-Metabolism p 95
108. Joel Fuhrman – Super Immunity p 111
109. Natasha Campbell-McBride – Gut & Psych Syndrome pp 11-13
110. Ibid p15

111. Marc David – The Slow Down Diet pp 71,72,75
112. Natasha Campbell-McBride – Gut & Psych Syndrome p 15
113. Mark Hyman – Ultra Prevention p 270, 2003
114. Natasha Campbell-McBride – Gut & Psych Syndrome p 21
115. Neal D Barnard – The Cheese Trap p 49, 2017
116. Mark Hyman – Ultra Metabolism p 603
117. Natasha Campbell-McBride – Gut & Psych Syndrome p 21
118. Phyllis A Balch – Nutritional Healing p 28
119. Joel Fuhrman – Super Immunity p 89
120. Natasha Campbell-McBride – Gut & Psych Syndrome p 27
121. Ibid p 28
122. Mark Hyman – Change your brain, Change your life p 156
123. Mariza Snyder & Lauren Clum – The ORAC scale pp 9-12, 2011
124. Ibid p 14
125. Ibid p 17
126. Wendell Fowler – Eat Right Now! 2.0 p 101, 2014

Chapter 5 & Conclusions

127. Marc David – The Slow Down Diet p 3
128. Mark Hyman – Ultra Metabolism p 68
129. Byron J Richards – Fight for Your Health p 1, 2006
130. Michael Moss – Salt, Sugar, Fat p 213
131. Mark Hyman - Ultra Metabolism p 377
132. Jane Hawley Stevens – Web article, Organic foods
133. Edward Loeser – Clean up Your diet
134. Rudolph Ballentine – Radical Healing p 21
135. Robert J Thiel – Naturopathy 21st Cent. P 1
136. Daniel G Amen – Change your Brain, change your Life p 377
137. Mark Hyman – Ultra Metabolism p 80
138. Yan Ping Xu – on line article, sleep

About The Author

Dr. Marvin H. Massey Sr, ND, PhD is a naturopathic health practitioner and health coach. Naturopaths do not diagnose diseases or prescribe drugs.

Naturopaths suggest lifestyle changes and recommend proper dietary adjustments.

He earned the rank of Eagle Scout, and is a veteran where he served as a paratrooper in the U.S. Army 101st Airborne Division.

Dr. Massey began his journey as a healer in 1989 upon earning his 1st Degree Black Belt in Shorei-go Ryu Karate. Upon earning his 3rd Degree Black Belt in 1992, he was awarded his teaching certificate in karate, the sensei rank. In addition, to being a teacher a sensei must also become a healer, the Japanese translation of sensei is teacher or doctor, the theory is that if you harm someone you must be able to heal them as well. The historical first teacher of karate was Bodhidharma a Buddhist monk from India. He traveled over the Himalayan mountains into the country of China. He was an adept and healer in yoga.

It then occurred to Dr Massey that it would stand to reason that by studying yoga he could become a more proficient healer himself. So, he found a yoga instructor only three blocks from his house, taught by Barbra Jo Kennedy.

In 1995 he was led to study tai chi and Qi gong with instructor Leslie Mills, which are in the family of The Chinese healing arts. Traditional Chinese Medicine is based on the chi (energy) concept. Using a series of meridians in the body which are associated with the organs in the body. The healing points are used in acupuncture and acupressure.

He was introduced to the healing art of Reiki in 1999 and became a Reiki master and teacher. The following year he studied at the Academy of Reflexology and Heath Therapy International. There he received his certification in massage therapy and in that same year he became certified in reflexology. Dr Massey served as a member of the faculty at the same institution from 2006 through 2008.

He completed his B.S. in Holistic Health from the American Institute of Holistic Theology in 2010. In 2013 he received his M. H. in Herbology and the following year earned his Doctor of Naturopathy from Trinity College of Natural Health.

Dr Massey was awarded his 8th Degree Black Belt in 2017 by the late Grand Master

Herb Johnson.

The next year he earned his Doctor of Philosophy in Natural Medicine from The New Eden College of Clinical Natural Medicine.

He is an Ordained Minister in the Order of Melchizdek.

As of this writing he still actively teaches the martial arts and guides people to make healthy choices. One of his goals is to educate men, women and children in correct thinking, correct acting and correct eating. It is all about the choices we make and we choose the lifestyle that we are living.

Dissertation Bibliography-References

Amen, Daniel G MD, *Change Your Brain Change your Life*, 1998

Amen, Daniel G MD, *Change Your Brain Change your Body*, 2010

Balch, Phyllis A CNC, *Nutritional Healing*, 2010

Ballentine, Rudolph MD, *Diet and Nutrition*, 1978

Ballentine, Rudolph MD, *Radical Healing*, 2011

Barnard, Neal D MD, *The Cheese Trap*, 2017

Cameron, Ewan MB & Pauling, Linus PhD, *Cancer & Vitamin C*, 1979

Campbell-McBride, Natasha MD, *Gut & Psychology Syndrome*, 2010,

Chestnut, James L DC, *The Innate Diet & Natural Hygiene*, 2004

David, Marc, *The Slow Down Diet,* 2005

Duffy, William, *Sugar Blues*, 1975

Erasmus, Udo, *Fats that Heal, Fats that Kill*,1986

Fuhrman, Joel MD, *Super Immunity*, 2011

Fowler, Wendell Chef, *Eat Right Now 2.0, It's All About the Food*, 2014

Herbert, Victor MD & Subak-Sharpe, Genell MS, *Total Nutrition,* 1995

Hill, Napoleon, *Think & Grow Rich*, 1960

Hyman, Mark MD & Liponis, Mark MD, *Ultra prevention*, 2003

Hyman, Mark MD, *Ultra Metabolism*, 2006

Hyman, Mark MD, *Eat Fat, Get Thin*, 2016

Loomis, Howard F Jr DC, *Enzymes the Key to Health,* 2007

Magner, Lois N, *A History of Medicine,* 1992

Maurer, Robert PhD, *One Small Step Can Change Your Life, The Kaizen Way*, 2004

Moss, Michael, *Salt, Sugar, Fat*, 2013

Petrucciani, Lori ND, White, Bea, *Bring Back Vitality*, 2013

Richards, Byron J, *Fight for Your Health,* 2006

Snyder, Mariza DC, Clum Lauren DC, *The Antioxidant Counter,* 2011

Thiel, Robert J PhD, *Naturopathy for the 21st Century*, 2000

Wallach, Joel DVM.ND, Lan, Ma MD, Schrauzer, Gerhard N PhD, *Epigenetics*, 2014

Printed in the United States
By Bookmasters